A RETAILER'S GUIDE TO OSHA

A RETAILER'S GUIDE TO OSHA

George Matwes
Helen Matwes

021415

Chain Store Publishing Corporation
An Affiliate of Lebhar-Friedman, Inc.
New York

Copyright © 1976, Chain Store Publishing Corporation
425 Park Avenue, New York, N.Y. 10022

Printed in the United States of America
Library of Congress Catalog Card Number: 76–19179
International Standard Book Number: 0–912016–52–3

Dedication

To Arlen and Richard who cheered us from the sidelines.

In the multitude of counsellors there is safety.
Proverbs 11:14–15

Salus populi suprema lex
The people's safety is the highest law.
Maximus

Acknowledgements

We would like to express our sincere thanks to the following people for their help, suggestions, and interest in the preparation of this book.

Harry Demarest, Marsh & McLennan, Inc.

George Groves, Food Fair, Inc.

Leonard Laufer, Occupational Safety for Industry, Inc.

Kenneth Ryan, Zurich Insurance Company

And our special thanks to Alfred Barden, OSHA Regional Administrator, New York City, and James Conlon, OSHA Area Director, Belle Mead, New Jersey, who readily answered all our questions concerning OSHA's rules and regulations.

Helen and George Matwes

Introduction

This OSHA reference book is directed toward the retail industry. We feel we have chosen those standards that are most likely to affect the retailer in his day to day operations. This volume is meant as a guide; its contents should not be construed as an official and final directive since OSHA rules and regulations are in an ever-changing condition based on new concepts, appeals, and OSHA Review Commission decisions.

It can, however, be an excellent tool in helping to understand and comply with current OSHA standards. And it can be instrumental in improving safety programs which, in turn, will not only help in compliance with OSHA's requirements, but will also, in the long run, save costly insurance expense dollars.

Contents

021415

Chapter 1

Why OSHA?

In an important poll taken recently among retail supervisors, this question was asked: What are two of the most important things for which you, as a retail supervisor, are responsible? The answers added up to two basic items—sales volume and expense control.

These are not only the responsibilities of the supervisor, but also of any employee from the president on down. Anything which interferes with these two responsibilities must be of immediate concern, and steps must be taken to prevent any such interference.

The loss of an employee's time due to illness or absenteeism is one such interference. Another major interference which results in loss of time involves injury to people and/or damage to merchandise, machines, or equipment.

It is our belief that the government, through the Occupational Safety and Health Act (OSHA), has become a partner with the retailing executive and his staff to help control accidents to store employees and property. The control is brought about by getting people to do their jobs the way they should be done, safely and efficiently, in an established, safe workplace. Indeed, an unsafe workplace has now become a federal offense!

THE OCCUPATIONAL SAFETY
AND HEALTH ACT

In 1970, Senators Harrison A. Williams Jr. (D-N.J.) and William
A. Steiger (R-Wisc.) put together one of the most comprehensive
and far-reaching pieces of labor legislation ever passed by the U.S.
Congress. Called the "Williams-Steiger Occupational Safety and
Health Act," it became effective in April, 1971. The act affects
employers involved in almost every form of commerce and indus-
try, including retailing. All OSHA standards, rules, and regulations
apply to any retailer "in any state of the United States, the District
of Columbia, the Commonwealth of Puerto Rico, the Virgin Islands,
American Samoa, Guam, Trust Territory of the Pacific Islands, Wake
Island, Outer Continental Shelf Lands, Johnston Island, and the
Canal Zone."

The OSH Act requires the employer to furnish a safe workplace, free from recognized hazards that can cause, or are likely
to cause, death, permanent or partial injury, or any physical harm
to employees. It also requires that records be made of occupational
injuries and illnesses which will be used for further studies in con-
trolling these hazards. In addition, a special poster must be ex-
hibited outlining OSHA's rules and regulations.

CONCERNING CUSTOMERS

This book is based on the retailer's responsibilities under OSHA to
his employees. Although it does *not* concern itself basically with
customer safety, it is important to note that because employees go
wherever customers go in a retail establishment, to make the place
safe for employees automatically makes it safe for customers.

At one point, we raised the question with an OSHA representative as to what would happen if a customer wrote to OSHA
in Washington, D.C. and complained about an unsafe condition in
a store. The reply was, "Well, though she is not an employee, still,
it is a potential hazard and we would probably have one of our
Compliance Officers check it out."

Chapter 2

The OSHA Law

THE FEDERAL REGISTER

The Federal Register is the official publication of the U.S. Government, published daily Monday through Friday. It contains notices of meetings, hearings, proposed regulations, adapted regulations, and other legal notices. The Federal Register does *not* contain copies of the laws and statutes themselves.

The first publication of the OSHA law, which has since been revised, appeared in the Federal Register of Wednesday, October 18, 1972. As of this writing, the latest updated Federal Register publication for OSHA is dated Thursday, June 27, 1974. It can be obtained from the U.S. Government Printing Office, Washington, D.C. for 20 cents.

The Federal Register for OSHA is a voluminous book (each page contains three long columns of fine print) covering the OSHA rules and regulations. Since its length and details, headings, subheadings, and sub-sub-headings often make it difficult and time consuming to find the applicable OSHA regulations, it is the purpose of this book to select those standards that concern you, the retailer.

HOW TO USE
THE FEDERAL REGISTER

All OSHA standards have been assigned Part Number 1910 in the Federal Register. The rules and regulations are divided into lettered subparts or general subjects. The subjects are further divided, and each division is assigned a 1910 number plus a decimal number, as follows:

Subpart		Sections in the Federal Register
A	General	1910.1 through 1910.6
B	Adoption and Extension of Established Federal Standards	1910.11 through 1910.19
C	(Reserved)	(Reserved)
D	Walking-Working Surfaces	1910.21 through 1910.32
E	Means of Egress	1910.35 through 1910.40
F	Powered Platforms, Manlifts and Vehicle-Mounted Work Platforms	1910.66 through 1910.70
G	Occupational Health and Environmental Control	1910.93 through 1910.100
H	Hazardous Materials	1910.101 through 1910.116
I	Personal Protective Equipment	1910.132 through 1910.140
J	General Environmental Controls	1910.141 through 1910.149
K	Medical and First Aid	1910.151 through 1910.153
L	Fire Protection	1910.156 through 1910.165b
M	Compressed Gas and Compressed Air Equipment	1910.166 through 1910.171
N	Materials Handling and Storage	1910.176 through 1910.184
O	Machinery and Machine Guarding	1910.211 through 1910.222

P	Hand and Portable Powered Tools and Other Hand-Held Equipment	1910.241 through 1910.247
Q	Welding, Cutting, and Brazing	1910.251 through 1910.254
R	Special Industries	1910.261 through 1910.269
S	Electrical	1910.308 through 1910.309

Let's suppose you wish to learn what the standards are for Fixed Industrial Stairs which is listed under Subpart D, Walking-Working Surfaces. This Subpart is divided into numbered sections with Fixed Industrial Stairs numbered 1910.24. You turn to the back of the Federal Register to learn on which page 1910.24 appears. The back of the register, which is an alphabetical index, shows that Fixed Industrial Stairs appears on page 23510.

Another way to get information from the Federal Register is to use the alphabetical index first. You wish to look up Circular Saws, for instance. You would look for Woodworking Tools under which you would find that Circular Saws is numbered 1910.243 (a) (1) and appears on page 23734. On page 23734 we find that 1910.243 is titled "Guarding of Portable Powered Tools." The portion (a) is on portable powered tools and (1) refers to portable circular saws. (These pages are taken from Volume 39, Number 125 of the Federal Register dated Thursday, June 27, 1974.)

STANDARDS

Definition

A "standard" is a ruling which requires conditions, or the adoption of one or more practices, means, methods, operations, or processes reasonably necessary or appropriate for a safe workplace.

Petition for the Issuance, Amendment or Repeal of a Standard

Any interested person may petition (in writing) the Assistant Secretary of Labor to introduce, modify, or revoke a standard.

The request must state the basic rule and the change or addition recommended plus the reason for the recommendation.

Application of Standards

If a particular OSHA standard is specifically applicable to a working condition, it will take precedence over any general standard which might otherwise apply to the same condition. In other words, OSHA standards take precedence over other prescribed standards for all industries.

As an example, the general machine guarding requirements are contained in part 1910.212 of the Federal Register. In 1910.213 specific requirements are given for guarding woodworking machinery. For those types of saws specifically mentioned in 1910.213, such as radial saws, the specific requirements override the general requirements of 1910.212. However, when a type of woodworking equipment is not specifically named in 1910.213, the more general requirements of 1910.212 will apply.

Sources for Standards Used by OSHA

The OSHA safety standards are generally quite reasonable. They were developed from those already in effect, from the U.S. Departments of Labor and Health, Education, and Welfare, the National Electrical Association, the National Fire Protection Association (NFPA), the American National Standards Institute (ANSI), and many other sources.

Below is a list of addresses of some of these organizations.

American National Standards Institute (ANSI)
1430 Broadway
New York, N.Y. 10018

National Fire Protection Association (NFPA)
470 Atlantic Avenue
Boston, Mass.

American Society of Heating, Refrigeration, and Air
Conditioning Engineers, Inc. (ASHRAE)
345 E. 47th St.
New York, N.Y. 10017

U.S. Department of Health, Education, and
Welfare (HEW)
Washington, D.C. 20203

American Society of Mechanical Engineers
345 E. 47th St.
New York, N.Y. 10017

Underwriters Laboratories, Inc.
207 E. Ohio St.
Chicago, Illinois 60611

American Society of Safety Engineers (ASSE)
Park Ridge, Illinois

Note: The National Electrical Code is a national consensus standard. Write to the National Fire Protection Association (NFPA) for further details.

Chapter 3

OSHA:
The Government's
Responsibilities

Under the OSHA law, if you are engaged in a business that in any way affects interstate commerce—and that means just about any employer—you must provide a "safe and healthful place in which to work" and which is "free from any recognized hazards." In order to ensure that this is being done, the government must:

- Set safety standards.
- Issue citations for safety violations.
- Devise appropriate forms of warning to keep employees informed of hazards.
- Supervise OSHA standards through job-site inspections.
- Prescribe suitable protective equipment for employees.
- Issue guidelines to employees and employers in occupational sickness and injury control.
- Research new methods and techniques for dealing with problems of job safety and health.

The occupational safety and health standards, developed under the Occupational Safety and Health Act of 1971, are issued by the Department of Labor's Occupational Safety and Health Administration (OSHA). All OSHA standards are published in the Federal Register, as explained in Chapter 2.

The first OSHA standards, printed in the Federal Register of October 18, 1972, were based principally on the safety and health standards that had been in existence for several years through various professional and technical associations. In the ensuing two years, several of the standards were changed or altered to comply with court decisions and reinterpreted meanings that came up as the OSHA law was implemented across the country. OSHA requires that:

1. Employers and employees reduce hazards in the workplace and start or improve existing safety and health programs.

2. Employer and employee responsibilities be established.

3. Mandatory job safety and health standards be set.

4. An effective enforcement program be provided. Professionals who enforce the OSH Act are officially called Compliance Safety and Health Officers and Industrial Hygienists. These are men and women with years of experience and professional training. They receive a highly specialized training course before moving into the field to make inspections. At least once a year, they must take a refresher course plus additional training in specialized fields such as construction or maritime safety at OSHA's training institute in Rosemont, Illinois. Some training programs are conducted in the various area offices. An extensive on-the-job training program is given in all OSHA areas. When a Compliance Officer completes his training program he is still considered on a six-month training status and still under the close supervision of an experienced Compliance Officer.

5. The states assume the fullest responsibility for administering and enforcing their own occupational safety and health programs which are required to be at least as effective as the federal program.

    6.   An accident recording procedure concerning job injuries, occupational injuries, and fatalities be set up.

The Occupational Safety and Health Administration, arranges specific Retailer's Safety Courses to be given at designated colleges in all states. Information on these can be obtained through an OSHA Regional Office. All courses given under OSHA are free to all participants.

OSHA also conducts one-day seminars for employers. These are conducted frequently in each of the 51 areas across the nation. Again, the Regional Office is the main source of information on these seminars. Seminars can be arranged upon the request of one employer or a group of employers. If you provide the hall and the audience, OSHA will provide the speakers.

To further aid you, the following are some of OSHA's publications that will help keep you up to date on various OSHA activities, standards changes, Review Commission changes, etc.

Job Safety and Health is the official OSHA magazine. It keeps you up to date on standards, industry/union activities, job health research, current safety and health issues, news from the National Institute of Safety and Health (NIOSH),* and information from the OSHA Review Commission. It also lists current OSHA publications, including the latest Federal Register entries on the OSHA law. Write: Superintendent of Documents, Government Printing Office, Washington, D.C. 20402. ($13.60 per year)

Occupational Safety and Health Subscription Service is the best way to keep up with changes in the standards. This also includes a volume on other regulations, such as regulations on methods of inspection, recordkeeping regulations and so on. It also includes a volume on the Compliance Operations Manual which has now become the Field Operations Manual and which is in the process of being extensively revised. There also is a booklet on recordkeeping.

Occupational Safety and Health Standards Digest. This is a summary of the more important standards. The last revision is

---

* NIOSH (part of the Department of Health, Education, and Welfare) concerns itself with health-related aspects of the work place while OSHA is concerned with the purely physical safety of the work place.

dated March, 1975 and carries the number OSHA 2201. Single copies can be obtained from the Regional Office and it can be purchased in quantity from the Superintendent of Documents, at a price of $1.05. The forward in the digest says in part "although the digest does not contain all the general industry safety and health standards, those selected cover approximately 90 percent of the basic applicable standards and are expressed in simple terms. In addition, in the back of this booklet a standard source section is included, key to the numerical designations for each of the standards listed in the digest."

You also might refer to the *NIOSH Health and Safety Guides*. Single copies of these are available from the Office of Technical Publications, National Institute for Occupational Safety and Health, Post Office Building, Cincinnati, Ohio 45202. A self-addressed mailing label should be included with the request. There is a series of these guides and more are being published each month.

All federal regulations are printed in what is known as the *Code of Federal Regulations*, taking up many volumes. The Department of Labor Title 29 of the Code (29CFR) is printed in several volumes. The volumes are six by nine inches in size and paper bound. New editions are printed once a year on a staggered schedule. The last printing of Title 29 contained two volumes with OSHA regulations, Part 1900 through 1919 which sells for $7.35 and Part 1920 through the end which sells for $3.50. This really is the official code; the daily issues of the Federal Register supplement this by printing changes since the last printing of the regular code volumes.

ON-SITE CONSULTATIONS

One of the major employer complaints since the establishment of the OSHA law has been that the arrival of an unannounced Compliance Officer for an inspection automatically resulted in citations and fines. Many employers claimed they did not know the rules, exactly what was to be inspected, the necessary safeguarding standards, and other elements of the OSH Act.

Although ignorance of the law has never been accepted as an official excuse, the government has arranged for employers to have the assistance of consultants to examine the workplace for unsafe or hazardous conditions and to advise on the proper methods to alleviate such conditions. It must be remembered, however, that an on-site consultation must deal with specific problems and is not to be considered a "free" OSHA inspection.

In order for an employer to get a consultation, he must check with the area OSHA Director in his state. Many states have entered into consultation agreements between the state department of labor and/or industry and OSHA to provide these consultation services. If there is such an agreement, then the state, not OSHA, provides the inspection and consultation staff.

Purposes of On-Site Consultations

There are several purposes of this on-site consultation program:

1. To provide employers with the information necessary to comply with standards, rules, and regulations of OSHA.
2. To assist employers in understanding their obligations and responsibilities under the Act.
3. To consult and advise employers as to the effective means of preventing occupational injuries and illness.
4. To encourage employers to achieve and maintain safe and healthful workplaces.

OPERATION OF THE PROGRAM

A state which enters into a Consultation Agreement agrees to provide several qualified safety and/or industrial hygienist consultants. In general, in order for such consultants to qualify for this program, they must meet the requirements for state employment in the occupational safety and health field. They must have adequate education and experience in occupational safety and health to satisfy the Assistant Regional Director that they have the ability to perform satisfactorily.

All state consultants selected for this program undergo an OSHA four-week training course designed specifically for this program.

Consultations under this program will be provided *only upon request* of an employer, with the highest priority given to small businesses, and further consideration given to the hazardous nature of the employer's activities.

If an employer has a number of different working conditions in his workplace, he will be required to refer to the specific working conditions for which he seeks consultation, and consultation will be limited to those conditions specified. However, the smaller the business, the less specific the reference to working conditions will have to be; the larger the establishment, the more specific the reference. Another order of priority, but lower, will be given to follow-up visits, off-site consultation, education and training, and other voluntary compliance activities as approved by the Assistant Regional Director.

The purpose of on-site consultation is to advise on specific problems—*it is not* designed to provide a free OSHA inspection.

After an employer has requested a consultation visit, and the visit is scheduled, the consultation would proceed as follows:

Opening Conference

During the opening conference, the consultant would present his appropriate state credentials. Since there have been several OSHA imposters, it is important to check their credentials and, if there is still any doubt, to call the consultant's office to verify authenticity. The consultant will advise the employer that, in the event of a subsequent OSHA inspection, the Compliance Officer *will not* be legally bound to the advice given by the consultant or by the failure of the consultant to note a specific hazard.

The consultant will tell the employer that he *may*, but is not required to, furnish a copy of the consultation report to an inspecting Compliance Officer. The employer's refusal to provide a copy *will not* imply bad faith on his part. However, if the employer chooses to make the report available to the Compliance Officer, it may be used to determine the nature of violations and the employer's good faith or lack of good faith.

The employer must be advised by the consultant that, upon discovery of an apparent imminent danger situation, he will be requested to eliminate the condition immediately or, if this is not possible, to remove employees from the dangerous area. A follow-up visit is mandatory in situations where abatement of the hazard cannot be effected in the presence of the consultant. This is required in order to ensure that employees are not being re-exposed to a danger. If an employer fails to cooperate in eliminating a danger, the consultant must immediately advise the OSHA Assistant Regional Director of the situation.

Employees or their representatives *may* participate in the consultation visit, but only with the permission of the employer. The consultant will ascertain whether and under what circumstances he may confer with employees.

The Walk-Through

During his walk-through of the store, the consultant will explain to the employer which OSHA standards and regulations apply to his workplace. In addition, the consultant will look specifically at those items for which the consultation visit was requested. He will advise the employer if and how he is not in compliance with OSHA standards and regulations. Where it is feasible and within his technical competence to do so, the consultant will suggest the means by which identified hazards may be abated.

All information reported to or otherwise obtained by the state in connection with any consultation which contains or which might reveal a trade secret will be considered confidential.

Industrial hygienist consultation services are limited to the use of direct reading instruments only. In the event the industrial hygienist consultant identifies a potential health hazard which requires laboratory facilities, he will refer the employer to available sources for confirmation of the health hazard. Such sources may include, but are not limited to, other state services, private consultants, insurance carriers, and NIOSH (National Institute of Safety and Health).

The Closing Conference

During the closing conference of his visit, the consultant will summarize his findings and answer the employer's questions. The employer will receive a written report summarizing the consultation visit and the consultant's findings and recommendations.

The Report

As soon as possible, but no later than ten working days, the employer will be provided with a written report of the findings and recommendations.

The report will identify the specific hazards discovered, whether they were of a serious or non-serious nature, the number of instances of the hazards identified, their locations, a reference to the OSHA standard which applies, and a suggested means of eliminating the hazards. Where there is no specific applicable OSHA standard, the General Duty Clause, 5(a)(1), will be indicated.

Information related to a consultation visit will not trigger an OSHA inspection since the OSHA enforcement staff and the state consultation staff are completely independent of one another, and it is their policy not to utilize this program for enforcement purposes, except in cases of imminent danger not abated.

As with state plans, OSHA will monitor the performance of the consultants and the consultation program, and changes in procedures may be directed by the OSHA Regional Administrator within the scope of OSHA policy.

Chapter 4

OSHA:
The Employer's
Responsibilities

OSHA requirements concerning employers state: "Each employer shall furnish to each of his employees employment and a place of employment free from recognized hazards that are causing or are likely to cause death or serious harm to his employees; and shall comply with occupational safety and health standards issued under the Act."

Standards pertaining to retailers fall into three general classifications. The first classification includes the physical plant, parking lot, buildings, floors, stairs, electricity, heating and ventilation, fire protection, sanitary facilities, etc. The second classification concerns operational equipment such as tools, machine guards, toxic materials, personal protection equipment, etc. The third classification deals with the various training standards, safety activities of employees, and specialized OSHA record keeping.

Where a standard of your state's Department of Labor overlaps a federal OSHA standard, the employer must follow the stricter of the two. Any necessary clarifications should be requested of the regional OSHA office.

The enforcement of the OSH Act is accomplished through a series of inspections. If the employer is satisfied with the results of an inspection or if he does not object by filing a notice of contest, he will have to pay any specified fees for safety violations found on the inspection. If a notice of contest for a variance is filed, however, the matter is taken up by the OSHA Review Commission. Even after the Review Commission makes a decision on a request for variance, both the employer and the Secretary of Labor have the right to appeal the decision to the federal court system, usually starting with the Circuit Court of Appeals.

Presently, OSHA is faced with the task of inspecting about five million workplaces throughout the United States and its territories, but its staff is much too small for this job. At one time OSHA adopted a rule of inspecting the "worst first" using a priority system. After these priorities, workplaces were inspected on a basis of random selection.

However, as of this date, there are no inspection priorities and no special emphasis programs. Both the Target Industry Program and the Target Health Hazard Program have expired. There may be some new special emphasis programs established in the near future, but no firm information is available at this writing. Presently, most inspections seem to cover these categories:

- Imminent danger
- Catastrophe/fatality investigations
- Employee complaints
- Regional Programmed Inspection (used to be called Random Inspection).

We feel that the following reasons, in order, are the most likely to trigger an OSHA inspection of a *retail operation:*

- Imminent Danger
- Catastrophe or fatality
- Employee complaints
- Regional Programmed Inspection

When a Compliance Officer comes to your establishment to make an inspection, he looks for compliance with national safety and health standards. He is concerned with what standards apply to you as a retailer (both general and specific) and whether you and your employees are complying with them. The Compliance Officer uses a Field Operations Manual (formerly called Compliance Operations Manual) as a detailed guide in his inspections. This manual lists these categories of violations:

1. Imminent Danger: "Any conditions or practices which are such that a danger exists which could reasonably be expected to cause death or serious physical harm . . ."

2. Serious Condition: "If there is a substantial probability that death or serious physical harm could result from a condition that exists, or from one or more practices, etc."

3. Non-Serious Condition: "Where violation . . . would probably not cause death or serious physical harm, but which has a direct or immediate relationship to the safety or health of employees."

4. Willful Neglect of a Hazardous Condition

5. Repeated Neglect of a Hazardous Condition

Three factors used in determining the "gravity of violation" are:

1. The probability or likelihood of injury or disease taking place as a result of the alleged violation.

2. The severity of the injury or disease which is most likely to result.

3. The extent to which the applicable provision of the OSH Act was violated.

Each factor is rated "A" through "C" and an average is determined. Penalties are assessed on the following monetary scale:

A: None
B: $100–$200
C: $201–$500
X: $501–$1000

"X" is used only when the employer demonstrates a definite disregard for the violation in question. (See chapter on Citations for more details.)

Generally, before making an inspection of your premises, the Compliance Officer familiarizes himself with as many relevant facts as possible concerning safety of retail establishments and those OSHA standards which will be pertinent to his inspection.

Inspections of premises are conducted during regular working hours except in special circumstances. (Because of an employee complaint, the author once had an OSHA Compliance Officer complete his inspection of a store at 9:30 p.m. A complaint had been filed concerning a night noise factor. The Compliance Officer arrived during the day and continued with his inspection until closing time. The complaint was deemed unwarranted.)

OSHA regulations do not allow for an advance notice of inspections (except where such notice would serve to make the inspection more effective). Therefore, it is very important for you, as a retailer, to set up a planned procedure in preparation for an OSHA inspection. Here are some suggestions.

PREPARING FOR AN
OSHA COMPLIANCE OFFICER
INSPECTION

In order for the store management to be prepared to effectively handle an inspection by an OSHA representative, it is particularly important to create a favorable impression upon his arrival when he asks to see the "appropriate employer representative." The "appropriate employer representative" should be the store manager or, if he is not on the premises, either the store personnel manager or the store operations manager. If this is not feasible because of their absence, various alternate employees (all on the management level) should be pre-designated to serve in this capacity.

Everyone concerned should be briefed on the following plan

of action: All receptionists (or secretaries) working in the area where the Compliance Officer would initially present himself when he enters the store should be given careful instructions. The Compliance Officer is to be referred *immediately* to the store manager, or, in his absence, the next most important member of management. There should be a list of people, in the order of their importance, to which the receptionist can refer should top level management be absent from the premises. The person available will act as the initial "welcoming committee." Compliance Officers should not be left sitting in waiting rooms for any more than five or ten minutes, since they might consider this a form of evasion to "clean up" before they are allowed to inspect.

Let us assume the store manager is in. He should give full priority to having the Compliance Officer ushered into his office as quickly as possible. While the officer is to be treated with full respect, a request for and examination of his credentials should be made by the store manager to be certain that he is a bona fide representative of OSHA. Since there have been instances of individuals who have falsely represented themselves as Compliance Officers for OSHA, the manager should call the area OSHA office for verification if he has any doubt about the officer's credentials.

Assuming all is in order, the officer should now inform the store manager of the reason for his visit (an OSHA inspection, an employee complaint, etc.) and outline in general terms the scope of the inspection he plans. It will probably include all safety and health records for his review, any employee interviews, a walk around the store premises (back and front), and a closing conference. If the officer's visit is due to an employee's complaint, he will give the store manager a copy of the complaint. If the employee had requested that his name be withheld, the manager will be given only a typed copy of the complaint, with no signature.

After he has reviewed and inspected the complaint situation, the Compliance Officer has the right either to leave or to make a full inspection of the store even though his visit was originally prompted by a specific complaint.

Before taking a walk-around inspection of the store's premises, it would be a good idea for the store manager to call in the store personnel manager and operations manager and introduce them. We then suggest that either the store manager or whoever he designates accompany the Compliance Officer on his walk-around inspection. It also would be a good idea to request the company nurse or whoever keeps safety and health records for the store to bring copies of them into the office. The officer can and usually does inspect the following records:

- Log of Occupational Injuries and Illness (OSHA Form 100)
- Copy of Summary of Occupational Injuries (OSHA Form 102)

OSHA forms 100 and 102 must be up-to-date when requested by the Compliance Officer. The Compliance Officer should be shown other safety and health records which the store should be keeping. These records would include:

- Minutes and reports of the store's Safety Committee meetings.
- Reports of various safety inspections conducted by any of the store's safety personnel and/or the insurance company's inspections.
- Any safety indoctrination material: booklets, training films, slides for new employees, etc.
- Any specialized training films, slides, etc. where safety is emphasized (for the receiving, shipping, stock, or warehouse employees).
- Any articles, notes, points of information concerning safety and health that have been used in employee house organs, publications, handouts, etc. Any other pertinent data, records, etc. to show that the store maintains an active and ongoing safety program for all employees.

"Good Faith" Benefits

OSHA Compliance Officers, when making an inspection of a store, are asked to determine if the company has shown "good faith" in meeting the intentions of the OSH Act by having an ongoing store safety program which might include the following:

- Official store safety inspections
- Safety Committee meetings and minutes
- Accident reporting and follow-up procedures
- Either a full or a part-time employee assigned to safety

The Compliance Officer, when reviewing any citations with his regional officer, will take into account the "good faith" features of a store's safety program and accordingly may reduce the actual penalties by as much as 50 percent.

THE INSPECTION

Before leaving for the walk-around inspection, the Compliance Officer will ask if there is a union or unions in the store; if so, he is authorized to request that the employer arrange to have an employee representative join them. The employer does not select the employee representative. That is done by the employee's organization itself. If there are no employee groups, the Compliance Officer can discuss working conditions with individual employees as he walks around the premises.

(It is suggested that if your retail establishment has an ongoing Safety Committee and if there is a union that a shop steward of the union be on the committee. At the time of the inspection he can be designated as the employee representative. If your store has no union, then a member of the Safety Committee would make a good walk-around representative.)

Neither the employee representative nor management may harass or otherwise obstruct the Compliance Officer's inspection process. Again, let us stress the importance of having a top store executive accompany the Compliance Officer on his inspection. It is also suggested that the executive have a pad and pencil with him so that when the officer takes appropriate notes of unsafe or hazardous conditions, he can do the same. If the Compliance Officer takes any photographs, and he has a right to do so, it may be advisable for the store executive to do the same.

The store representative also should take notes on any comments concerning unsafe or hazardous conditions. Tape recorders are not recommended because they tend to inhibit a relaxed exchange of information which is the real goal of the visit.

It is extremely important that the company representative who accompanies the Compliance Officer make every attempt to have any unsafe condition or hazardous practice corrected at once. For example, a blocked aisle should be brought to the department manager's attention right away so that he can have it cleared immediately.

Such corrections are recorded by the Compliance Officer and help in judging employer "good faith" in compliance. Even though corrected, the apparent violation may be the basis for a citation or penalty. However, the "good faith" shown can affect the dollar amount of the penalty.

During the inspection the Compliance Officer may be stopped by any employee who wishes to bring to his attention any condition believed to be a hazard or unsafe practice.

As he makes his tour of inspection, the Compliance Officer will be checking some of the following items:

- Has the employer made the necessary efforts to remove any hazards that could cause injury?
- Has required safety equipment been made available to the employees—and are they using them?
- Have the plant, its equipment, and its tools been properly maintained to meet safety standards?
- Have the necessary warning and other signs, color codes, etc. been put into their proper locations?
- Have the OSHA standards specifically set up for that particular kind of establishment been put into effect?
- Have employees been trained in the necessary safety procedures for lifting, handling certain equipment, and other jobs?
- Has a reporting procedure for all employee accidents been set up according to OSHA requirements?

After the inspection, the officer should be taken back to the manager's office where he will conduct an "exit interview" or a closing conference. At this time the officer discusses with management what he has seen and reviews probable violations. The Compliance Officer may *not*, on his own, impose or propose a penalty "on the spot" *at the time of the inspection.*

It is recommended that the manager or store representative take notes on the officer's comments at this closing conference. The Compliance Officer will ask the manager how long he will need in order to abate any hazards. Giving estimated abatement time for each hazard is very important because in the official citation report the store can be penalized for any days over the set abatement date. If an employer cannot meet the abatement dates listed in the citation, he can contact the OSHA office for a Petition of Modification of Abatement (PMA). A copy of the PMA is on page 25.

The Compliance Officer then goes back to his office and discusses his report with the Area Director. The Area Director (or his superiors, if necessary) determine what citations will be issued and what penalties, if any, will be proposed. All citations must refer to a specific violation of a specific section of an OSHA standard.

The store will then receive the citation report by certified mail. It is important to make arrangements with your secretarial staff that any certified mail from the Occupational Safety and Health Administration must be given to you, personally, at the time of arrival. If you are not in the store at that time, it must be given to the personnel manager or the operations manager or another designated store executive. This is particularly important because you have only 15 working days to appeal any citation, abatement period, or penalty. Therefore, every day counts.

If contested, OSHA violations are subject to final action by a separate authority, the Occupational Safety and Health Review Commission which will be discussed later in the book.

After the store has received the official citation report, the manager can call the nearest OSHA Regional Office (see locations in the Appendix of this book) and ask the OSHA Area Director for an informal hearing to discuss the possibility of amending (removing) one or more of the citations, giving reasons why he does not believe they are applicable. If he is not questioning their applicability, the manager can request more abatement time or a more

§ 1903.14a Petitions for modification of abatement date.

(a) An employer may file a petition for modification of abatement date when he has made a good faith effort to comply with the abatement requirements of a citation, but such abatement has not been completed because of factors beyond his reasonable control.

(b) A petition—for modification of abatement date shall be in writing and shall include the following information:

(1) All steps taken by the employer, and the dates of such action, in an effort to achieve compliance during the prescribed abatement period.

(2) The specific additional abatement time necessary in order to achieve compliance.

(3) The reasons such additional time is necessary, including the unavailability of professional or technical personnel or of materials and equipment, or because necessary construction or alteration of facilities cannot be completed by the original abatement date.

(4) All available interim steps being taken to safeguard the employees against the cited hazard during the abatement period.

(5) A certification that a copy of the petition has been posted and, if appropriate, served on the authorized representative of affected employees, in accordance with subsection (c)(1) of this section and a certification of the date upon which such posting and service was made.

(c) A petition for modification of abatement date shall be filed with the Area Director of the United States Department of Labor who issued the citation no later than the close of the next working day following the date on which abatement was originally required. A later-filed petition shall be accompanied by the employer's statement of exceptional circumstances explaining the delay.

(1) A copy of such petition shall be posted in a conspicuous place where all affected employees will have notice thereof or near such location where the violation occurred. The petition shall remain posted for a period of ten (10) working days. Where affected employees are represented by an authorized representative, said representative shall be served with a copy of such petition.

(2) Affected employees or their representatives may file an objection in writing to such petition with the aforesaid Area Director. Failure to file such objection within ten (10) working days of the date of posting of such petition or of service upon an authorized representative shall constitute a waiver of any further right to object to said petition.

(3) The Secretary or his duly authorized agent shall have the authority to approve any petition for modification of abatement date filed pursuant to paragraphs (b) and (c) of this section. Such uncontested petitions shall become final orders pursuant to sections 10 (a) and (c) of the Act.

(4) The Secretary or his authorized representative shall not exercise his approval power until the expiration of fifteen (15) working days from the date the petition was posted or served pursuant to paragraphs (c) (1) and (2) of this section by the employer.

(d) Where any petition is objected to by the Secretary or affected employees, the petition, citation, and any objections shall be forwarded to the Commission within three (3) working days after the expiration of the fifteen (15) day period set out in paragraph (c)(1) of this section.

This amendment shall be effective on February 12, 1975.

(Secs. 8(g)(2), 10(c), 84 Stat. 1600, 1601 (29 U.S.C. 657(g)(2), 659(c)))

Signed at Washington, D.C. this 5th day of February 1975.

JOHN STENDER,
Assistant Secretary of Labor.

[FR Doc.75-3796 Filed 2-10-75; 8:45 am]

021415

detailed statement as to how the Administrator wants the store to remove the causes that brought about the citations.

The store representative can request this informal hearing by telephone or mail within ten days after notification of the citation. This means that if, after the hearing, the store still wishes to do so, it can file officially (within the extra five days still left of the 15 allowed days) for a "notice of contest."

We have learned that requests for informal hearings on citations have found the OSHA Area Director very cooperative and responsive to arguments.

The informal hearing is very important. It is an opportunity for the employer to meet face to face with the Area Director and the Compliance Officer involved to discuss the citation and penalty. Remember, the Area Director or the Regional Administrator has the authority to:

- Amend or revoke the citation.
- Amend or revoke the penalty.
- Modify the abatement date.

Once the opportunity for an informal conference is passed and a notice of contest is filed, the Area Director no longer has the authority to make changes.

Chapter 5

OSHA:
The Employee's
Responsibilities

As stated previously, the employer under OSHA has a "general duty" to provide a place of employment free from recognized hazards and to comply with all the standards of the OSH Act. If he does not do this he can be fined (cited) and even have his business closed down until it is made safe to work in.

On the other hand, even though he or she is specifically required to comply with all safety and health standards that apply to actions and conduct on the job and in the workplace, the employee is not subject to any government sanctions when company safety regulations are not followed. Disciplinary action is at the option of the employer; he can treat a safety violation in the same manner as he would treat any violation of an established company rule or procedure.

Although OSHA does not prescribe penalties for employees who do not follow company safety rules and regulations, failure to obey them should precipitate reprimand, suspension, or dismissal from the job if the violation is serious enough.

Here is a checklist for your employee to follow. He or she should:

- Read the OSHA poster which you have displayed on the bulletin board.
- Comply with all OSHA standards that apply to the job.
- Follow all the employer's safety and health rules and regulations and wear or use all prescribed equipment and gear furnished by the employer.
- Report all unsafe conditions or hazardous practices to the supervisor or someone designated by the employer.
- Report all job-related injuries or occupational illnesses to the employer to be given prompt treatment.
- Cooperate with the OSHA Compliance Officer if the latter inquires about conditions at the work place.
- Exercise rights under OSHA in a responsible manner, by:
  - Not making safety devices inoperative
  - Reporting unsafe fixtures, equipment, tools, etc.
  - Serving on the Store Safety Committee when requested
  - Participating fully in all company fire drill activities

The following is a list of 29 rights to which employees are entitled according to OSHA:

1. The right to be furnished a place of employment free from recognized hazards that are causing or likely to cause death or serious physical harm to employees.
2. The right to request that a rule pertaining to a particular standard be determined by the Secretary of Labor.
3. The right to be informed of any standard which may affect employees.
4. The right to be advised of any order issued by the Secretary of Labor granting a variance to an employer.

5. The right to be informed when an application for a variance might be requested.

6. The right to petition the Secretary of Labor for a hearing pertaining to any variance request.

7. The right to be informed by labels or other appropriate forms of warning of all hazards to which they may be exposed.

8. The right to be furnished suitable protective equipment.

9. The right to have locations monitored in order to measure employee's exposure in such a manner as may be necessary for their protection.

10. The right to be given medical examinations or other tests to determine whether employee health is being affected by an exposure.

11. The right to have results of examinations or tests furnished to their physician.

12. The right to request modification or revocation of a variance allowed by the Secretary of Labor.

13. The right to be questioned privately by a federal Compliance Officer as he inspects the workplace.

14. The right to have regulations posted to inform employees of the protection afforded under OSHA.

15. The right to have NIOSH (National Institute of Safety and Health), a division of the Federal Department of Health, Education and Welfare (HEW), monitor or measure possible exposure to toxic materials.

16. The right to have access to the records of such monitoring or measuring.

17. The right to have a record of their own personal exposure.

18. The right to be promptly notified of any exposure to toxic materials.

19. The right to be informed of corrective action being taken to eliminate exposure.

20. The right of an employee representative to accompany

the federal Compliance Officer during the inspection of any work place.

21. The right to request an inspection if there may exist exposure to physical harm or if an imminent danger exists. Such a request may be made by an employee with the right not to have his name divulged to his employer.

22. The right to be advised formally by the Secretary of Labor or his designee of the determination that there are no reasonable grounds for such an inspection.

23. The right to advise a federal Compliance Officer of any violation of the OSH Act which he believes exists in the workplace.

24. The right to be informed of citations made to the employer by having them promptly posted.

25. The right to file with the Secretary of Labor a belief that the abatement time fixed in the citation for correction is unreasonable and be given an opportunity to participate in hearings which may be held pertaining to the citation.

26. The right not to be discriminated against by the employer because of any complaint filed or inspection requested. This right includes the opportunity to testify.

27. The right not to have variances in effect for more than six months without being notified by the Secretary of Labor.

28. The right to obtain a copy of the OSHA standards and other rules, regulations, and requirements from the employer or the nearest OSHA or U.S. Printing Office.

29. The right to request information from the employer on any safety and health hazards that exist in the workplace, on the necessary precautions he must take, on protective equipment he needs to wear, and on what he must do if he is involved in an accident or exposed to toxic substances.

Generally, the OSH Act seems designed to extend the rights to employees to be informed consistently before, during, and after

OSHA inspections. It indicates that an employee requesting either an OSHA or NIOSH inspection must be specific and name the hazards that concern him, but that he should first make an effort to have the employer correct the condition before notifying either OSHA or NIOSH. There always will be some disgruntled employees who wish to harass a company, but OSHA will not automatically make an inspection each time it gets a request. However, it will make an inspection when it feels there are reasonable grounds to believe a violation or a danger exists or if the request is in writing and from a bona fide employee or his authorized representative.

Chapter 6

Citations, Variances, and Appeals

Now comes the moment of truth. The OSHA Compliance Officer has been through your establishment. He has told you both the good news of how well you have met the OSHA standards and the bad news of the violations of the standards he has noted. He now tells you that, "with reasonable promptness," you will be receiving from the OSHA regional office, by certified mail, a Citation document and with it, a "Notification of Proposed Penalties."

The exact definition of a "Citation" (see page 33) is a notification, in writing, of the precise nature of the alleged violation, including a reference to the provision of the act, standard, rule, regulation, or order alleged to have been violated. It will state a reasonable period of time (generally agreed to at your exit interview with the Compliance Officer) for the correction of the alleged violation. This time is called the "abatement period."

The "Notification of the Proposed Penalty" (see page 34) for each violation is the amount in dollars you are fined for that violation.

If neither the cited employer nor any affected employees take action to contest these citations *within 15 working days* (work-

	CSHO NO.	OSHA-1 NO.	FY
	AREA	REGION	

Certified Mail

CITATION

U.S. DEPARTMENT OF LABOR
OCCUPATIONAL SAFETY AND HEALTH ADM.

RETURN RECEIPT REQUESTED

TO: 2.

3. Citation Number_____

4. Page _____ of 5. _____

6.
TYPE OF ALLEGED VIOLATION(S):

7.
An inspection was made on_____ 19 ___ of a place of employment located at:

8.
9. _____ and described as follows:

On the basis of the inspection it is alleged that you have violated the Occupational Safety and Health Act of 1970, 29 U.S.C. 651 *et seq.*, in the following respects:

10. Item number	11. Standard, regulation or section of the Act allegedly violated	12. Description of alleged violation	13. Date by which alleged violation must be corrected

The law requires that a copy of this citation shall be prominently posted in a conspicuous place at or near each place that an alleged violation referred to in the citation occurred. The citation must remain posted until all alleged violations cited therein are corrected, or for 3 working days*, whichever period is longer.

RIGHTS OF EMPLOYEES

Any employee or representative of employees who believes that any period of time fixed in this citation for the correction of a violation is unreasonable has the right to contest such time for correction by submitting a letter to the U.S. Department of Labor at the address shown above within 15 working days* of the issuance of this citation.

"No person shall discharge or in any manner discriminate against any employee because such employee has filed any complaint or instituted or caused to be instituted any proceeding under or related to this Act or has testified or is about to testify in such proceeding or because of the exercise by such employee on behalf of himself or others of any right afforded by this Act." Sec. 11(c) (1) of the Occupational Safety and Health Act of 1970, 29 U.S.C. 651, 660(c)(1).

*Under the Occupational Safety and Health Act, the term "Working Day" means Mondays through Fridays but does not include Saturdays, Sundays, or Federal Holidays.

14.
Area Director's Signature _____ Issuance Date _____ 19 ____

(NOTICE: Additional Important Information On Reverse Side)

Form OSHA-2

U. S. DEPARTMENT OF LABOR
Occupational Safety and Health Administration

⌐ 1. ¬

☐ New   ☐ Amended

2. CSHO NO.	3. OSHA - NO.
4. AREA	5. REGION

L   ⌐

TO:
6.

7. Page _____ of _____

8. Date _____

NOTIFICATION OF PROPOSED PENALTY

THERE IS NO REQUIREMENT THAT THIS NOTIFICATION BE POSTED

This notification and the penalty(ies) proposed by the Secretary of Labor shall be deemed to be the final order of the Occupational Safety and Health Review Commission (*an independent agency with authority to issue decisions respecting citations and proposed penalties*) and not subject to review by any court or agency <u>unless</u>, within 15 working days from the date of receipt of this notification, you submit a letter of contest. The letter of contest should be mailed or otherwise delivered to the Area Director named below at the address shown at the top of this notification. If no notice of contest is filed within the 15 working day period the proposed penalty(ies) becomes final and is immediately payable.

Payment of all penalties shown is to be made <u>by check or money order</u> payable to the order of "Occupational Safety and Health-Labor". Payment of penalties should be remitted to the Area Director at the address shown above.

Section 17(1) of the Act states: "Civil penalties owed under this Act shall be paid into the Treasury of the United States and shall accrue to the United States and may be recovered in a civil action in the name of the United States brought in the United States district court for the district where the violation is alleged to have occurred or where the employer has its principal office."

9.

On the _____ day of _____ , a citation(s) was issued to you in accordance with the provisions of Section 9(a) of the Occupational Safety and Health Act of 1970 (84 Stat. 1601; 29 U.S.C. 651, <u>et seq.</u>) hereinafter referred to as the Act. You were thus notified of certain alleged violations of the Act, as specified in that citation(s).

YOU ARE HEREBY NOTIFIED that pursuant to the provisions of Section 10(a) of the Act, the penalty(ies) set forth below is/are being proposed, based on the citation(s).

10. Citation Number	11. Item Number	12. Standard, Regulation or Section of the Act allegedly violated	13. Date by which alleged violation must be corrected	14. Proposed Penalty

15. Area Director	Date	16. Total Proposed Penalty for all alleged violations.	$

Rev. 8/75 OSHA-3

ing days are Mondays through Fridays, excluding federal holidays),
the fines must be paid. Payment is addressed to the attention of the
area regional office that issued the citations. After 15 days have
elapsed and if the citations were not contested, they will no longer
be subject to review by any court or agency.

NOTE: A recent review claim concerned the mailing of a
Certified Mail "Citation" and "Notification of Penalty" to a plant
superintendent. The superintendent, fearing for his job, threw the
citation letter away. When the company finally was cited again,
this time very heavily, and it learned by investigation what had
happened, it appealed and won. Because of this case, the rule now
is that "Citations" and "Notification of Penalties" must be sent to
an officer of the company or a special individual designated by the
plant or store manager. This person should be designated at the
exit interview with the inspecting Compliance Officer.

AMENDING
AN OSHA CITATION
VIA INFORMAL VARIANCE

Any retailer not necessarily wishing to request a formal variance
can, within ten days, call or write the OSHA Area Director for an
informal conference. At this conference the employer will either
receive clarification of the citation or learn how to go about correct-
ing it to meet OSHA standards. The employer might ask for an
amendment (variance) of the citation, based on reasons he gives
for believing that he should not have been cited in the first place.

CONTESTING AN OSHA CITATION
VIA FORMAL VARIANCE
OR NOTICE OF CONTEST *

If the retailer feels that the citation is totally unfair and unwarranted and is willing to request a variance formally, then:

* A "variance" is an official approval to follow a different standard than that
imposed on others. A "notice of contest" is an appeal from a citation. Area
Directors are not involved in requests for variances. These are handled directly
by the national office. However, the Director can be asked for an informal
conference concerning the appeal from a citation.

- Within 15 working days from his receipt of the notification of the proposed penalty, he must notify the U.S. Department of Labor, in writing, of his intention to contest (this is called a "Notice of Contest"). If the retailer has a legal department, obviously it should be consulted first. If the company has no legal department, it is advisable, although not mandatory, to consult an attorney who is well versed in labor relations.

- Upon receiving notice that the case has been scheduled, the retailer must notify his affected employees (and their union representative, if any) that he is contesting the citation.

If the retailer receives, for example, seven citations and wishes to contest only items 2, 4, and 6, he should so specify and pay the monetary penalties on the non-contested items.

A decision to contest should not be made lightly. Once a notice of contest is filed, it is a matter before the court, in this case the Review Commission. The only way a notice of contest can be withdrawn is by agreement between the Secretary of Labor and the employer. If an employer files a notice of contest and then fails to reply to the various legal papers served upon him, it is possible for the employer to be found guilty of willful or repeated violation.

The variance hearings will take place in front of an Occupational Safety and Health Review Commission judge. The proceedings are adversary proceedings, conducted like trials in a court. Each party to the proceedings may call witnesses, introduce evidence, and cross-examine opposing witnesses. The burden of proof is on the Secretary of Labor, through his agent, the OSHA Compliance Officer. When the hearing is over and before the judge files his decision, each party is given the opportunity to submit written briefs on their side of the case.

The judge then renders his written decision and files it with the full Review Commission. This becomes effective 30 days after its receipt by the Commission unless, within those 30 days any Commission members direct that the case be reviewed by the Commission. The full Commission, based on its findings, issues an order

HOW TO GET A VARIANCE

The Occupational Safety & Health Act authorizes the granting of variances from OSHA Standards to employers who are unable to comply with them — or who provide equally safe and healthful conditions for their employees.

OSHA's Office of Standards has prepared a **"Model Application For Variance"** reprinted below. *Note that the application letter:*

(1) Contains the applicant's name and place of employment; (2) details the alternative system in use; (3) sets forth the conditions which make installation of the indicator as required by OSHA Standards unfeasible; and (4) explains why the alternative method is more effective.

The variance application must be signed by a responsible company representative — and contain a certification that a copy was posted for employees to read. The posted copy must be accompanied by an explanation of the employees' right to comment on the application.

The Model also contains an application for an interim order permitting continued operation at variance with the OSHA Standard until a decision is made on the application.

Assistant Secretary for Occupational Safety and Health
U.S. Department of Labor
Washington, D.C. 20210

Dear Sir:

Pursuant to Section 6 (d) of the Williams-Steiger Occupational Safety and Health Act of 1970 (84 Stat. 1596: 29 USC .655), respectfully requests a permanent variance from the requirements of 29 C.F.R. 1910.179 (b) (4), concerning wind indicator on the outdoor bridge of a Gantry Crane.

(1) Applicant
(2) Place of employment - Same as (1)
(3) has a Gantry Crane which operates beside a slip for unloading shallow draft barges. It currently is used only for unloading salt barges, usually about twice a year. Because of the location of our operation and the experience we have had with high wind conditions, recognizes the importance for monitoring wind conditions. As a result, our plant has two wind indicating devices with visual and automatic recording devices. We have a well established plan for monitoring Federal Weather Bureau reports and emergency plans of action for pending high winds. The key wind indicating device is located in the Shift Superintendent's office. A Shift Superintendent is on duty around the clock, seven days a week and is responsible for the plant. He keeps a close watch on the wind conditions and also monitors weather forecasts at least three times during each eight-hour period. A written record is kept of the weather conditions. If there are any changes in wind conditions or anticipated changes due to approaching bad weather, the wind indicators and weather forecast are monitored continuously. Because of these well established policies and the limited use of the Gantry, we feel that our plan would offer a safer operation. In addition, we feel that the installation of a wind indicating device on the Gantry Crane would not be feasible because of conditions which tend to cause malfunction. Our past experience has verified this consistently. The Gantry Crane is very close to the molten sulfur pits and salt water. The close proximity to these agents creates a corrosive condition which would make the wind indicator inoperative soon after installation. We feel our monitoring and early warning system is more effective and safe than

is the use of an infrequently used and difficult to maintain wind velocity instrument on the crane.

(4) Due to the corrosive conditions present around the Gantry Crane, feels that our established procedures would give more reliable information than an indicator on the crane. Therefore, we feel that this would insure the safety of the crane and operator better by following our procedures.

Respectfully submitted,

Manager

(5) This is to certify that a copy of the variance from Standard 1910.179 (b) 4 requested by has been posted in an appropriate place for employees to examine. Also, an explanation of their right to send comments or petition for a hearing are explained on an attached sheet. A copy of the variance application was also given to the authorized employee representative.

Posted on _____ , 19

Manager

(6)   **TO ALL EMPLOYEES**

The attached form is a copy of the variance application submitted by for a variance from Standard 1910.179 (b) 4. This is to inform you that you may request a hearing with the Assistant Secretary for Occupational Safety and Health, U.S. Department of Labor, according to Standard 1905.15, Request for Hearing, if you object to the variance requested. This request must be filed in writing.

Manager

(7) also respectfully requests that an interim order be granted to allow operation of the Gantry Crane without the indicator until action is taken on the variance application. We feel that the safety of the employees will be guaranteed because we do have wind indicators being monitored at our plant and the Gantry operator would be promptly informed before wind velocities rose or approached a dangerous level.

Manager

affirming, modifying, or vacating the citation. The decision, whatever it is, becomes final in 30 days after the hearing. An appeal from the decision can be filed with the appropriate U.S. Circuit Court of Appeals. See procedure How to Get A Variance and the Model Application for Variance on page 37.

Chapter 7

OSHA Standards

The following are the most important OSHA standards concerning the retailer. (Many of these are discussed in greater detail elsewhere in this book.) Some of these standards are open to technical and legal interpretation. For instance, what is a "permanent" aisle? How "promptly" must a spill be cleaned up? What does "near proximity" mean—one mile, three miles, etc., etc.? It is not the intention of the authors to go into these interpretations, but rather to explain the standard as written in the law in language understandable to the lay person.

WALKING AND WORKING
SURFACES

- All passageways, storerooms, and maintenance shops must be maintained clean, dry, orderly, and in a sanitary condition. Spills must be cleaned up promptly.

- Areas which are constantly wet should have non-slip surfaces where personnel normally work.
- Every floor, working place, and passageway must be maintained free from protruding nails, splinters, holes, loose boards, and, as far as possible, in a dry condition.
- Where mechanical handling equipment such as lift trucks are used, clearances must be provided for aisles, loading docks, through doorways, and wherever turns or passage must be made. No obstructions that could create a hazard are permitted in the aisles.
- All permanent aisles must be easily recognizable by the use of painted or otherwise marked aisle lines.
- No exit, aisleway, or passageway should be obstructed or blocked.

FIXED LADDERS

- Must have rungs with a minimum diameter of three-quarter inch for metal ladders or one and one-eighth inches for wood ladders.
- Must not have rungs spaced more than 12 inches apart; rungs must be at least 16 inches wide.
- Must be equipped with "safety cages" if they are longer than 20 feet.

PORTABLE LADDERS

- Must be maintained in good condition and inspected frequently. Defective ladders must be tagged "Dangerous—Do Not Use" and removed from service for repair or destruction.
- Must not be placed on boxes or barrels or other supports to obtain additional height.

ROLLING LADDERS

- Must have working "brakes" to anchor ladder. Legs must have rubber tips.
- Must not have too much sway at top of ladder.

FIXED INDUSTRIAL STAIRS

- Rise height and tread must be uniform.
- All treads must be slip-resistant.
- Vertical clearance above any stair tread to an overhead obstruction must be at least seven feet.
- All stairs should be adequately lighted.
- If the flight of stairs has four or more risers, railings or handrails must be provided.
- If a stairway is less than 44 inches wide and open on both sides, a stair railing on each side is required.
- If the stairway is less than 44 inches wide and open on one side, a stair railing on the open side is required.
- If both sides are enclosed on a stairway less than 44 inches wide, at least one handrail is required, preferably on the right side descending.
- If the stairway is more than 44 inches wide but less than 88 inches wide, a stair railing on each open side and a handrail on each enclosed side are required.
- If each stairway is 88 inches wide or more, a handrail on each enclosed side, a stair railing on each open side, and an additional stair railing located midway are required.
- The vertical height of the railing must be between 30 and 34 inches, and the railing must be smooth surfaced.
- A standard railing consists of a top rail, intermediate rail, and posts. The distance from the upper surface of

the top rail to the floor platform runway or ramp must be 42 inches. The intermediate rail must be approximately halfway between the top rail and the floor.

- A standard toeboard must be approximately four inches in height from the floor to the top edge with no more than one-quarter inch gap between the toeboard and the floor. As a general condition, a standard toeboard and railing are required wherever people walk beneath the open sides of a platform, because things could fall from such a structure.

MEANS OF EGRESS

- Every building or structure, new or old, designed for human occupancy shall be provided with exits sufficient to permit the prompt escape of occupants in case of fire or other emergency.
- All building or structure exits shall be so arranged and maintained to provide free and unobstructed egress from all parts of the building or structure.
- No lock or fastening to prevent free escape from the inside of any building shall be installed except in mental, penal, or corrective institutions. (There are approved security-type locks on doors which, in an emergency, can be opened from the inside by just pushing against a small bar which automatically opens the door. These are referred to as "panic bars.")
- Every approved fire exit must have the word *EXIT* in legible lettering not less than six inches high.
- Doors, passageways, or stairways which are neither exits nor lead to exits should be clearly indicated by a sign reading, "Not An Exit" or by a sign designating its actual use.
- When the direction to the nearest exit is not apparent, an exit sign with an arrow indicating direction where an exit exists must be used.

- Nothing impairing the visibility of the exit sign, such as decorations, furnishings, or other signs, should be allowed.

- All exit doors must be side-hinged swinging types. They must swing out in the direction of travel.

- If occupancy is permitted at night, exit signs must be suitably illuminated by a reliable light source.

- In every building equipped for artificial illumination, all exits must have adequate and reliable illumination.

FIRE PROTECTION
(See Warehousing)

ELECTRICAL EQUIPMENT

- Attachment plugs, connectors, receptacles, etc. for use on portable equipment must be of the grounding type.

- Attachment plugs and connectors must be rated to the size of the equipment and designed to prevent insertion into a receptacle of lower rating.

- Equipment is acceptable if it is accepted, certified, listed, labeled, or otherwise determined to be safe by a nationally recognized testing laboratory, such as, but not limited to, Underwriters' Laboratories, Inc. and Factory Mutual Engineering Corporation.

- Every new electrical installation and all new utilization equipment installed after March 15, 1972 and every replacement, modification, or repair or rehabilitation after March 15, 1972 of any part of any electrical installation or utilization equipment installed before March 15, 1972 must be installed or made and maintained in accordance with the provisions of the 1971 National Electrical Code (National Fire Protection Association 70–1971) and the American National Standards Institute (ANSI CI-1971, revision of 1968).

- Every means of disconnection, circuit breakers, fuse box, etc., must be legibly marked to indicate its purpose unless its purpose is evident.
- Areas where wires are joined, such as outlets, switches, junction boxes, etc. must be covered.
- Parts of electrical equipment which, in ordinary use, arc or spark, must be enclosed.
- Flexible cords may not be used:
  - As substitutes for fixed wiring.
  - When run through holes in walls, ceilings, or floors.
- Flexible cords must be:
  - Continuous lengths without splices or taping.
  - Fastened so that there is no pull on joints or screws.
  - Replaced when frayed.
  - Grounded when equipment is connected by flexible cords.

OCCUPATIONAL HEALTH AND ENVIRONMENTAL CONTROLS

Retailers may find the following recommendations important if, in their various operations, they use the following:

- *Asbestos.* (Used for repair of brakes in auto centers.) Where brake repairs go on daily or where linings of brakes are machined to fit drums, excessive asbestos exposure could exist. Operators should be required to wear dust masks. Dust should be vacuumed (not blown) from the drums and the floor vacuumed, not swept.
- *Carbon Monoxide.* Any concentration of automobile or truck exhaust (not diesel) indoors can be a source of carbon monoxide. (Use of general ventilation, tailpipe, exhaust systems, and adequate mechanical ventilation are ways to combat this problem. Inspect your mechani-

cal exhaust systems, frequently checking for blockage with paper and rags, fan belt slipping or breaking, or hose breaking or splitting.)

- *Solvents.* Cleaning tanks should contain local exhaust ventilation to the outside.

- *Paints and Thinners*
  - Spray booths should be mechanically ventilated.
  - Spray painters should wear respirators.
  - Painters should not be allowed to wash their hands in the thinners.
  - Storage paints and thinners should be stored in well ventilated and approved storage cabinets or vaults.

- *Sanitation*
  - Safe drinking water must be provided.
  - Receptacles for waste food should be kept covered and clean.
  - Restrooms should be clean, provided with soap and water, toilet paper, and, in female lavatories, sanitary napkin machines.
  - No food or beverage is to be stored in a toilet area.

PERSONAL PROTECTIVE EQUIPMENT

Eye protection (safety goggles or face shields), hard hats, safety shoes, respirators, gloves, etc. must be worn as needed.

In many instances where the employee is required to work outdoors or even indoors but still subject to weather conditions, foul and/or cold weather gear should be supplied.

MEDICAL AND FIRST AID

All retailers must have on their premises an assigned First Aider (a nurse, either RN or LPN is preferred, of course) with an up-to-date

Red Cross Standard First Aid card, Bureau of Mines First Aid card, or other official certification. A first aid room should be supplied with at least a reclining chair and screen for any employees who need to be off their feet. First aid supplies, approved by the retailer's assigned physician, should be kept in the first aid room.

Chapter 8

Recordkeeping

THE REQUIRED RECORDS

The following are required by OSHA:

- OSHA Form No. 100—The Log of Occupational Injuries and Illness Record. To be completed within six workdays after employer learns of occupational injury or illness.

- OSHA Form No. 101—Supplementary Record of Occupational Injuries, the individual record of each illness or injury. To be completed and present in the establishment within six workdays after learning of injury or illness.

- OSHA Form No. 102—Summary of Occupational Injuries and Illness for Calendar Year. To be completed no later than one month after the calendar year. It must be posted no later than February 1 and not removed until after March 1. It is then returned to the file for future OSHA inspections.

(Note: You may substitute your own forms for OSHA Forms No. 100 and 101 as long as they contain all the information that appears on the OSHA forms and are as easily understandable and readable. Even a plain piece of paper will do. However, you must use the actual OSHA Form 102. If you choose to use your own forms for No. 100 and No. 101, we suggest you obtain copies to be able to identify codes and other pertinent information you will be obliged to incorporate for your records.)

These completed forms are not to be submitted as reports to the government. They are records to be kept current and on file, readily available for inspection and copying by the U.S. Department of Labor, Department of Health, Education, and Welfare, and your state Department of Labor.

Recordkeeping forms will not be mailed automatically to employers. Requests must be made either by phone or in writing to your OSHA regional or area office or state offices. (See list of phone numbers and addresses in the Appendix of this book.)

Any accident which results in one or more deaths or hospitalization of five or more employees must be reported within 48 hours, either orally or in writing, to the Area Director of the Occupational Safety and Health Administration, except in those states with approved state plans in which case report is to be made to the appropriate state agency. For Area Directors see telephone numbers and addresses under OSHA Regional Area Offices and for state agencies see Statistical Grant Agencies, both in the Appendix of this book. Further information may be obtained from the OSHA Area Director.

Who Must Keep Records

Most, but not all, private employers are required to keep records. Those who are exempt are small employers who employed no more than 11 full or part-time employees at any one time during the previous calendar year. (State health and safety laws however, may require records.) The only exception might be if a small employer is selected to participate in an annual survey of occupational injuries and illnesses. They will, however, be notified in advance and supplied with necessary forms.

If an employer operates several stores with a total of more than 11 employees, records must be kept for *each individual store*.

Where Records Are To Be Kept

Records are to be located where the business is conducted or operations performed. In other words, each workplace must have its own records on the premises.

Although other records must be maintained at the establishment to which they refer, you may prepare and maintain the log of injuries at another location (as in the case of data processing), but keep a copy updated to within 45 calendar days at all times on the premises of the establishment(s) where the injuries or illnesses occurred.

How Long Records Must Be Kept

All records must remain in the establishment and must be kept current for a period of five years after the end of the year to which they refer. The word "current" is important, for changes can occur since the original entry. For instance, if an employee at first required only medical treatment, but later lost workdays, the record would have to be changed. The entry in Column 10 in Form 102 (Non-Fatal Cases Without Lost Workdays) would be crossed out and an entry would be made in Column 9 and 9A (Lost Workday Cases and Number of Days Away from Work).

Another example of keeping the log current would be the case of an employee who lost workdays, returned to work, and then died of the illness. The lines 9, 9A, and/or 9B would be crossed out and the date of death entered in Column 8 (Deaths). Where a case formerly regarded as a work-related injury or illness is later determined to be non-work related, the entire entry would be crossed out. An entry in which it was thought medical treatment was given when actually only first aid was rendered would also be crossed out.

If an establishment changes ownership, the new employer is required to preserve the records for the remainder of the five years, but need not keep them current.

HOW TO PREPARE
AND MAINTAIN RECORDS

OSHA forms should be looked over carefully and completely before you begin; you will find they are not complicated. Again, there are three basic forms that must be maintained: the Log OSHA Form No. 100, the Supplementary OSHA Form No. 101, and the Summary OSHA Form No. 102.

The three recordable occupational injuries and illnesses are:

- Occupational Death
- Occupational Illness
- Occupational Injury

Detailed definitions of each of the above appear on the back of Form No. 100.

THE LOG (OSHA NO. 100)

The Log classifies the injury and illness cases and the extent and out-come of each one. This does not mean that every injury must be recorded. For instance, cases which require only first aid or cases which are not work-related need not be recorded. Any occupational injury or illness must be recorded within six workdays after learning of its occurrence.

Conditions resulting from animal bites such as dog bites which occur on the work premises are considered to be recordable injuries. Conditions resulting from one-time exposure to chemicals are considered occupational injuries and also are recordable.

You may substitute your own forms for OSHA No. 100, but the format must remain the same. For example, if you use data proc-essing equipment to log injuries and illnesses, you may use the print-out as long as it conforms to the same format as OSHA No. 100. However, it is best to obtain copies of Form No. 100 so that, if necessary, you can refer to specific definitions should questions arise.

To help you in preparing your Log, the following is an explanation of the information requested for each column appearing on the Log Form No. 100.

- Column 1—Case or File Number
  Use any series of non-duplicating numbers. Use same number on the Log as on the Supplement (Form No. 101 or its facsimile). This will act as a handy cross-reference. In other words, the file number on the Supplementary record should agree with the same case on the Log.

- Column 2—Date of Injury or Illness
  For occupational injury, record date of the accident in which the injury occurred. For an occupational illness, enter the date of the initial diagnosis. If absence from work occurred before illness was diagnosed, enter the first day of absence due to the illness which was later identified.

- Column 3—Employee's Name
  First name, middle initial, and last name.

- Column 4—Occupation
  Enter the job title, not the activity engaged in at the time of the injury or illness. If there is no official title, outline the duties normally performed by the employee.

- Column 5—Department
  The department where employee is *regularly* employed —even if this employee was temporarily working in another department at the time of injury or illness.

- Column 6—Nature of Injury or Illness and Parts of Body Affected
  Indicate parts of body affected. If entire body is affected, the word "body" is acceptable.

- Column 7—Injury or Illness Code
  On the bottom of Form No. 100 you will find the following Code Key which will give you the code number identifying an illness or injury.

CODE KEY

Code 10—All occupational injuries

Code 21—Occupational skin diseases or disorders

Code 22—Dust diseases of the lungs (Pneumoconioses)

Code 23—Respiratory conditions due to toxic agents

Code 24—Poisoning (systemic effects of toxic materials)

Code 25—Disorders due to physical agents (other than toxic materials)

Code 26—Disorders associated with repeated trauma

Code 29—All other occupational illness

(Remember, you are only to record work-related [occupational] injuries and illnesses. If there is a question as to which code applies in certain cases, there are detailed definitions on the back of the Log, OSHA Form No. 100, which will help you make your determination.)

- Column 8—Deaths
  Only deaths due to occupational injuries or illness. Enter date.

- Column 9—Lost Workdays
  Enter checkmark. Note that every entry for lost workdays will require an additional entry in 9A or 9B or both.

- Column 9A—Lost Workdays (Days Away From Work)
  Enter the number of workdays lost, even if they did not follow one another. Record only those days lost that the employee would *normally* have worked. Do not include the day of injury or the onset of the illness or any days the employee would not normally have worked. For part-time or part-week employees, it may be necessary to estimate lost workdays. Estimates of lost workdays should be based on the prior work history of the employee *AND* days worked by other employees (not ill or injured) who work in the same department or occupation of the injured or ill employee.

- Column 9B—Lost Workdays (Days of Restricted Work Activity)
  Enter number of lost workdays whether or not they

follow each other, where the employee because of ill-ness or injury:

1. Was assigned to another job on a temporary basis, or
2. Worked at a permanent job less than full time, or
3. Worked at a permanently assigned job but could not perform all the duties connected with it.

Do not include day of injury or onset of illness in the number of lost days. Also, do not include any days the employee would *not* normally have worked, had he been able.

● Column 10—Non-Fatal Cases Without Lost Workdays
Enter checkmark. This column will probably be the one you use most frequently, i.e. injury or illness that does not involve any lost workdays or fatalities.

● Column 11—Termination or Permanent Transfers
Enter checkmark if entries in columns 9 or 10 ended in termination or permanent transfer of employee.

THE SUPPLEMENTARY RECORD
(OSHA NO. 101)

This form need not necessarily be used if your Workmen's Compensation forms or other record forms contain all items listed on the OSHA No. 101 form. Missing information may be added to your own record forms and this will be acceptable. A plain piece of paper with all the necessary information is also acceptable.

There is a problem, however, in substituting Workmen's Compensation forms in place of the No. 101 in that the criteria for recording Workmen's Compensation differs from OSHA record-keeping and reporting requirements. Workmen's Compensation reg-ulations may require more or fewer cases to be recorded than OSHA's. For instance, in some states the Workmen's Compensation requires an injury to be reported if it results in at least two lost workdays. OSHA requires that even one lost workday be recorded. There are often other differences; therefore if you elect to use your

OSHA NO. 100

LOG OF OCCUPATIONAL INJURIES AND ILLNESSES

CASE OR FILE NUMBER	DATE OF INJURY OR ONSET OF ILLNESS	EMPLOYEE'S NAME (First name or initial, middle initial, last name)	OCCUPATION (Enter regular job title, not activity employee was performing when injured or at onset of illness.)	DEPARTMENT (Enter department in which the employee is regularly employed.)
(1)	Mo./day/yr. (2)	(3)	(4)	(5)

Company Name _____

Establishment Name _____

Establishment Address _____

NOTE: This is NOT a report form. Keep it in the establishment for 5 years.

☆GPO 615-686

54

RECORDABLE CASES: You are required to record information about: every occupational _death_; every nonfatal occupational _illness_; and those nonfatal occupational _injuries_ which involve one or more of the following: loss of consciousness, restriction of work or motion, transfer to another job, or medical treatment (other than first aid). More complete definitions appear on the other side of this form.

Form Approved
OMB No. 44R 1453

DESCRIPTION OF INJURY OR ILLNESS			EXTENT OF AND OUTCOME OF CASES					
			LOST WORKDAY CASES			NONFATAL CASES WITHOUT LOST WORKDAYS (Enter a check if no entry was made in columns 8 or 9 but the case is recordable, as defined above.)	TERMINATIONS OR PERMANENT TRANSFERS (Enter a check if the entry in columns 9 or 10 represented a termination or permanent transfer.)
					LOST WORKDAYS			
Nature of Injury or Illness and Part(s) of Body Affected (Typical entries for this column might be: Amputation of 1st joint right forefinger Strain of lower back Contact dermatitis on both hands Electrocution—body)	Injury or Illness Code See codes at bottom of page.	DEATHS (Enter date of death.)	Enter a check if case involved lost workdays.	Enter number of days AWAY FROM WORK due to injury or illness.	Enter number of days of RESTRICTED WORK ACTIVITY due to injury or illness.		
		Mo./day/yr.						
(6)	(7)	(8)	(9)	(9A)	(9B)	(10)	(11)

Injury Code

10 All occupational injuries

Illness Codes

21 Occupational skin diseases or disorders
22 Dust diseases of the lungs (pneumoconioses)
23 Respiratory conditions due to toxic agents
24 Poisoning (systemic effects of toxic materials)

25 Disorders due to physical agents (other than toxic materials)
26 Disorders associated with repeated trauma
29 All other occupational illnesses

LOG OF OCCUPATIONAL INJURIES AND ILLNESSES

Each employer who is subject to the recordkeeping requirements of the Occupational Safety and Health Act of 1970 must maintain for each establishment a log of all recordable occupational injuries and illnesses. This form (OSHA No. 100) may be used for that purpose. A substitute for the OSHA No. 100 is acceptable if it is as detailed, easily readable and understandable as the OSHA No. 100.

Each recordable occupational injury and occupational illness must be timely entered on the log. Logs must be kept current and retained for five (5) years following the end of the calendar year to which they relate. Logs must be available (normally at the establishment) for inspection and copying by representatives of the Department of Labor, or the Department of Health, Education and Welfare, or States accorded jurisdiction under the Act.

INSTRUCTIONS FOR COMPLETING LOG OF OCCUPATIONAL INJURIES AND ILLNESSES

Column 1 - CASE OR FILE NUMBER

Enter a number which will facilitate comparison with supplementary records. Any series of nonduplicating numbers may be used.

Column 2 - DATE OF INJURY OR ONSET OF ILLNESS

For occupational injuries enter the date of the work accident which resulted in injury. For occupational illnesses enter the date of initial diagnosis of illness, or, if absence from work occurred before diagnosis, enter the first day of the absence attributable to the illness which was later diagnosed or recognized.

Column 3 - EMPLOYEE'S NAME

Column 4 - OCCUPATION

Enter regular job title, not the specific activity being performed at time of injury or illness. In the absence of a formal occupational title, enter a brief description of the duties of the employee.

Column 5 - DEPARTMENT

Enter the name of the department or division in which the injured person is regularly employed, even though temporarily working in another department at the time of injury or illness. In the absence of formal department titles, enter a brief description of normal workplace to which employee is assigned.

Column 6 - NATURE OF INJURY OR ILLNESS AND PART(S) OF BODY AFFECTED

Enter a brief description of the injury or illness and indicate the part or parts of body affected. Where entire body is affected, the entry "body" can be used.

Column 7 - INJURY OR ILLNESS CODE

Enter the one code which most accurately describes the case. A list of the codes appears at the bottom of the log. A more complete description of recordable occupational injuries and illnesses appears in "DEFINITIONS."

Column 8 - DEATHS

If the occupational injury or illness resulted in death, enter date of death.

Column 9 - LOST WORKDAY CASES

Enter a check for each case which involves days away from work, or days of restricted work activity, or both. Each lost workday case also requires an entry in column 9A or column 9B, or both.

Column 9A - LOST WORKDAYS–DAYS AWAY FROM WORK

Enter the number of workdays (consecutive or not) on which the employee would have worked but could not because of occupational injury or illness. The number of lost workdays should not include the day of injury or onset of illness or any days on which the employee would not have worked even though able to work.

NOTE: For employees not having a regularly scheduled shift., i.e., certain truck drivers, construction workers, farm labor, casual labor, part-time employees, etc., it may be necessary to estimate the number of lost workdays. Estimates of lost workdays shall be based on prior work history of the employee AND days worked by employees, not ill or injured, working in the department and/or occupation of the ill or injured employee.

Column 9B - LOST WORKDAYS–DAYS OF RESTRICTED WORK ACTIVITY

Enter the number of workdays (consecutive or not) on which because of injury or illness:
1) the employee was assigned to another job on a temporary basis, or
2) the employee worked at a permanent job less than full time, or
3) the employee worked at a permanently assigned job but could not perform all duties normally connected with it.

The number of lost workdays should not include the day of injury or onset of illness or any days on which the employee would not have worked even though able to work.

Column 10 - NONFATAL CASES WITHOUT LOST WORKDAYS

Enter a check for any recordable case which does not involve a fatality or lost workdays.

Column 11 - TERMINATIONS OR PERMANENT TRANSFERS

Enter a check if the entry in columns 9 or 10 represented a termination of employment or permanent transfer.

CHANGES IN EXTENT OF OR OUTCOME OF INJURY OR ILLNESS

If, during the 5-year period the log must be retained, there is a change in a case which affects entries in columns 9 or 10, the first entry should be lined out and a new entry made. For example, if an injured employee at first required only medical treatment but later lost workdays, the check in column 10 should be lined out, a check entered in column 9, and the number of lost workdays entered in columns 9A and/or 9B.

In another example, if an employee with an occupational illness lost workdays, returned to work, and then died of the illness, the entries in columns 9, 9A, and/or 9B should be lined out and the date of death entered in column 8.

The entire entry for a case should be lined out if the case is later found to be nonrecordable. Examples are: A case which is later determined not to be work related, or a case which was initially thought to involve medical treatment but later was determined to have involved only first aid.

DEFINITIONS

RECORDABLE OCCUPATIONAL INJURIES AND ILLNESSES are:

1) OCCUPATIONAL DEATHS, regardless of the time between injury and death, or the length of the illness; or

2) OCCUPATIONAL ILLNESSES; or

3) OCCUPATIONAL INJURIES which involve one or more of the following: loss of consciousness, restriction of work or motion, transfer to another job, or medical treatment (other than first aid.)

NOTE: Any case which involves lost workdays must be recorded since it always involves one or more of the criteria for recordability.

OCCUPATIONAL INJURY is any injury such as a cut, fracture, sprain, amputation, etc., which results from a work accident or from an exposure involving a single incident in the work environment.
NOTE: Conditions resulting from animal bites, such as insect or snake bites, or from one-time exposure to chemicals are considered to be injuries.

OCCUPATIONAL ILLNESS of an employee is any abnormal condition or disorder, other than one resulting from an occupational injury, caused by exposure to environmental factors associated with employment. It includes acute and chronic illnesses or diseases which may be caused by inhalation, absorption, ingestion, or direct contact.

The following listing gives the categories of occupational illnesses and disorders that will be utilized for the purpose of classifying recordable illnesses. The identifying codes are those to be used in column 7 of the log. For purposes of information, examples of each category are given. These are typical examples, however, and are not to be considered to be the complete listing of the types of illnesses and disorders that are to be counted under each category.

(21) Occupational Skin Diseases or Disorders
Examples: Contact dermatitis, eczema, or rash caused by primary irritants and sensitizers or poisonous plants; oil acne; chrome ulcers; chemical burns or inflammations; etc.

(22) Dust Diseases of the Lungs (Pneumoconioses)
Examples: Silicosis, asbestosis, coal worker's pneumoconiosis, byssinosis, and other pneumoconioses.

(23) Respiratory Conditions Due to Toxic Agents
Examples: Pneumonitis, pharyngitis, rhinitis or acute congestion due to chemicals, dusts, gases, or fumes; farmer's lung; etc.

(24) Poisoning (Systemic Effects of Toxic Materials)
Examples: Poisoning by lead, mercury, cadmium, arsenic, or other metals, poisoning by carbon monoxide, hydrogen sulfide or other gases; poisoning by benzol, carbon tetrachloride, or other

organic solvents; poisoning by insecticide sprays such as parathion, lead arsenate; poisoning by other chemicals such as formaldehyde, plastics and resins; etc.

(25) Disorders Due to Physical Agents (Other Than Toxic Materials)
Examples: Heatstroke, sunstroke, heat exhaustion and other effects of environmental heat; freezing, frostbite and effects of exposure to low temperatures; caisson disease; effects of ionizing radiation (isotopes, X-rays, radium); effects of nonionizing radiation (welding flash, ultraviolet rays, microwaves, sunburn); etc.

(26) Disorders Associated With Repeated Trauma
Examples: Noise-induced hearing loss; synovitis, tenosynovitis, and bursitis; Raynaud's phenomena; and other conditions due to repeated motion, vibration or pressure.

(29) All Other Occupational Illnesses
Examples: Anthrax, brucellosis, infectious hepatitis, malignant and benign tumors, food poisoning, histoplasmosis, coccidioidomycosis, etc.

MEDICAL TREATMENT includes treatment (other than first aid) administered by a physician or by registered professional personnel under the standing orders of a physician. Medical treatment does NOT include first aid treatment (one-time treatment and subsequent observation of minor scratches, cuts, burns, splinters, and so forth, which do not ordinarily require medical care) even though provided by a physician or registered professional personnel.

ESTABLISHMENT: A single physical location where business is conducted or where services or industrial operations are performed (for example: a factory, mill, store, hotel, restaurant, movie theater, farm, ranch, bank, sales office, warehouse, or central administrative office). Where distinctly separate activities are performed at a single physical location (such as contract construction activities operated from the same physical location as a lumber yard), each activity shall be treated as a separate establishment.

For firms engaged in activities such as agriculture, construction, transportation, communications, and electric, gas and sanitary services, which may be physically dispersed, records may be maintained at a place to which employees report each day.

Records for personnel who do not primarily report or work at a single establishment, such as traveling salesmen, technicians, engineers, etc., shall be maintained at the location from which they are paid or the base from which personnel operate to carry out their activities.

WORK ENVIRONMENT is comprised of the physical location, equipment, materials processed or used, and the kinds of operations performed by an employee in the performance of his work, whether on or off the employer's premises.

Form approved
OMB No. 44R 1453

Supplementary Record of Occupational Injuries and Illnesses

EMPLOYER

1. Name _____

2. Mail address _____
      (No. and street)   (City or town)   (State)

3. Location, if different from mail address _____

INJURED OR ILL EMPLOYEE

4. Name _____ Social Security No. _____
      (First name)   (Middle name)   (Last name)

5. Home address _____
      (No. and street)   (City or town)   (State)

6. Age _____   7. Sex: Male_____ Female_____ (Check one)

8. Occupation _____
      (Enter regular job title, *not* the specific activity he was performing at time of injury.)

9. Department _____
      (Enter name of department or division in which the injured person is regularly employed, even
      though he may have been temporarily working in another department at the time of injury.)

THE ACCIDENT OR EXPOSURE TO OCCUPATIONAL ILLNESS

10. Place of accident or exposure _____
      (No. and street)   (City or town)   (State)

   If accident or exposure occurred on employer's premises, give address of plant or establishment in which
   it occurred. Do not indicate department or division within the plant or establishment. If accident occurred outside employer's premises at an identifiable address, give that address. If it occurred on a public highway or at any other place which cannot be identified by number and street, please provide place
   references locating the place of injury as accurately as possible.

11. Was place of accident or exposure on employer's premises? _____ (Yes or No)

12. What was the employee doing when injured? _____
                              (Be specific. If he was using tools or equipment or handling material,
   _____
                        name them and tell what he was doing with them.)
   _____

13. How did the accident occur? _____
                              (Describe fully the events which resulted in the injury or occupational illness. Tell what
   _____
   happened and how it happened. Name any objects or substances involved and tell how they were involved. Give
   _____
   full details on all factors which led or contributed to the accident. Use separate sheet for additional space.)

OCCUPATIONAL INJURY OR OCCUPATIONAL ILLNESS

14. Describe the injury or illness in detail and indicate the part of body affected. _____
                                                      (e.g.: amputation of right index finger
   _____
            at second joint; fracture of ribs; lead poisoning; dermatitis of left hand, etc.)

15. Name the object or substance which directly injured the employee. (For example, the machine or thing
      he struck against or which struck him; the vapor or poison he inhaled or swallowed; the chemical or ra-
      diation which irritated his skin; or in cases of strains, hernias, etc., the thing he was lifting, pulling, etc.)
   _____
   _____

16. Date of injury or initial diagnosis of occupational illness _____
                                                      (Date)

17. Did employee die? _____ (Yes or No)

OTHER

18. Name and address of physician _____

19. If hospitalized, name and address of hospital _____
   _____

   Date of report _____ Prepared by _____
   Official position _____

SUPPLEMENTARY RECORD OF
OCCUPATIONAL INJURIES
AND ILLNESSES

To supplement the Log of Occupational Injuries and Illnesses (OSHA No. 100), each establishment must maintain a record of each recordable occupational injury or illness. Workmen's compensation, insurance, or other reports are acceptable as records if they contain all facts listed below or are supplemented to do so. If no suitable report is made for other purposes, this form (OSHA No. 101) may be used or the necessary facts can be listed on a separate plain sheet of paper. These records must also be available in the establishment without delay and at reasonable times for examination by representatives of the Department of Labor and the Department of Health, Education and Welfare, and States accorded jurisdiction under the Act. The records must be maintained for a period of not less than five years following the end of the calendar year to which they relate.

Such records must contain at least the following facts:

1) *About the employer*—name, mail address, and location if different from mail address.

2) *About the injured or ill employee*—name, social security number, home address, age, sex, occupation, and department.

3) *About the accident or exposure to occupational illness*—place of accident or exposure, whether it was on employer's premises, what the employee was doing when injured, and how the accident occurred.

4) *About the occupational injury or illness*—description of the injury or illness, including part of body affected; name of the object or substance which directly injured the employee; and date of injury or diagnosis of illness.

5) *Other*—name and address of physician; if hospitalized, name and address of hospital; date of report; and name and position of person preparing the report.

SEE *DEFINITIONS* ON THE BACK OF OSHA FORM 100.

OSHA No. 102

Complete no later than one month after close of calendar year. See back of this form for posting requirements and instructions.

Form Approved
OMB No. 44R 1453

SUMMARY
OF
OCCUPATIONAL INJURIES AND ILLNESSES
FOR CALENDAR YEAR 19 __

Use previous edition of this form for summarizing your 1974 cases. This edition is for summarizing your cases for 1975 and subsequent years.

Establishment:
NAME _____
ADDRESS _____

INJURY AND ILLNESS CATEGORY		TOTAL CASES	DEATHS	LOST WORKDAY CASES				NONFATAL CASES WITHOUT LOST WORKDAYS	TERMINATIONS OR PERMANENT TRANSFERS
				Total Lost Workday Cases	Cases Involving Days Away From Work	Days Away From Work	Days of Restricted Work Activity		
CATEGORY	CODE	Number of entries in Col. 7 of the log. (1)	Number of entries in Col. 8 of the log. (2)	Number of checks in Col. 9 of the log. (3)	Number of entries in Col. 9A of the log. (4)	Sum of entries in Col. 9A of the log. (5)	Sum of entries in Col. 9B of the log. (6)	Number of checks in Col. 10 of the log. (7)	Number of checks in Col. 11 of the log. (8)
OCCUPATIONAL INJURIES	10								
Occupational Skin Diseases or Disorders	21								
Dust Diseases of the Lungs	22								
Respiratory Conditions Due to Toxic Agents	23								
Poisoning (Systemic Effects of Toxic Materials)	24								
Disorders Due to Physical Agents	25								
Disorders Associated With Repeated Trauma	26								
All Other Occupational Illnesses	29								
TOTAL—OCCUPATIONAL ILLNESSES (Sum of codes 21 through code 29)	30								
TOTAL—OCCUPATIONAL INJURIES AND ILLNESSES (Sum of code 10 and code 30)	31								

This is NOT a report form. Keep it in the establishment for 5 years.

I certify that this Summary of Occupational Injuries and Illnesses is true and complete, to the best of my knowledge.

Signature _____
Title _____
Date _____

SUMMARY OF OCCUPATIONAL INJURIES AND ILLNESSES

Every employer who is subject to the recordkeeping requirements of the Occupational Safety and Health Act of 1970 must use this form to prepare an annual summary of the occupational injury and illness experience of the employees in each of his establishments within one month following the end of each year.

POSTING REQUIREMENTS: A copy or copies of the summary must be posted at each establishment in the place or places where notices to employees are customarily posted. This summary must be posted no later than February 1 and must remain in place until March 1.

INSTRUCTIONS for completing this form: All entries must be summarized from the log (OSHA No. 100) or its equivalent. Before preparing this summary, review the log to be sure that entries are correct and each case is included in only one of the following classes: deaths (date in column 8), lost workday cases (check in column 9), or nonfatal cases without lost workdays (check in column 10). If an employee's loss of workdays is continuing at the time the summary is being made, estimate the number of future workdays he will lose and add that estimate to the workdays he has already lost and include this total in the summary. No further entries are to be made with respect to such cases in the next year's summary.

Occupational injuries and the seven categories of occupational illnesses are to be summarized separately. Identify each case by the code in column 7 of the log of occupational injuries and illnesses.

The summary from the log is made as follows:

A. For occupational injuries (identified by a code 10 in column 7 of the log form) make entries on the line for code 10 of this form.

    Column 1—Total Cases. Count the number of entries which have a code 10 in column 7 of the log. Enter this total in column 1 of this form. This is the total of occupational injuries for the year.
    Column 2—Deaths. Count the number of entries (date of death) for occupational injuries in column 8 of the log.
    Column 3—Total Lost Workday Cases. Count the number of checks for occupational injuries in column 9 of the log.
    Column 4—Cases Involving Days Away From Work. Count the number of entries for occupational injuries in column 9A of the log.
    Column 5—Days Away From Work. Add the entries (total days away) for occupational injuries in column 9A of the log.
    Column 6—Days of Restricted Work Activity. Add the entries (total of such days) for occupational injuries in column 9B of the log.
    Column 7—Nonfatal Cases Without Lost Workdays. Count the number of checks for occupational injuries in column 10 of the log.
    Column 8—Terminations or Permanent Transfers. Count the number of checks for occupational injuries in column 11 of the log.
    CHECK: If the totals for code 10 have been entered correctly, the sum of columns 2, 3, and 7 will equal the number entered in column 1.

B. Follow the same procedure for each illness code, entering the totals on the appropriate line of this form.

C. Add the entries for codes 21 through 29 in each column for occupational illnesses and enter totals on the line for code 30.

D. Add the entries for codes 10 and 30 in each column and enter totals on the line for code 31.

    CHECK: If the summary has been made correctly, the entry in column 1 of the total line (code 31) of this form will equal the total number of cases on the log.

The person responsible for the preparation of the summary shall certify that it is true and complete by signing the statement on the form.

GPO 956-893

Workmen's Compensation forms in place of the No. 101's, keep in mind that the OSHA criteria are to be followed.

If you use your own forms or records in place of Supplementary Record Form No. 101, they must include the following information:

- Employer
  Name, mailing address, and location if different from mailing address.

- The Injured or Ill Employee
  Name, Social Security number, home address, age, sex, occupation, and department.

- Description of Illness of Injury
  Full description of injury or illness, including part of body affected, name of object or substance which directly injured employee, and date of injury or diagnosis of illness.

- Other Pertinent Information
  Name and address of physician; if hospitalized, name and address of hospital; date of report; and name and position of person preparing report.

The Supplementary Record Form must be completed and present in the establishment six workdays after the employer has been notified of illness or injury.

THE SUMMARY OSHA NO. 102

This record is a summary of information which appears on the Log OSHA Form No. 100. It must be completed within one month after the end of each calendar year, and a copy or copies must be posted in each establishment by February 1 and remain posted until March 1. After March it should be placed back in files for future inspections. The posted Form 102 must be located where all employees will see it. If any employees on your February payroll do not report to the establishment on a regular basis, a copy must either be presented or mailed to each of them.

This is the one record that must appear on the OSHA Form No. 102 itself. No substitutions are acceptable.

Before preparing the summary, review the Log OSHA Form No. 100 to be certain it is correct and that each case is listed in only one of the following classes:

- Column 8: Deaths (would have a date)
- Column 9: Lost Workdays Cases (would have a check)
- Column 10: Non-Fatal Cases Without Lost Workdays (would have a check)

If an employee's loss of workdays is continuing at the time the summary is being made up, estimate the future workdays he will lose, add that estimate to the workdays he has already lost, and include this total in the summary. No further entries on this case need appear on your next year's summary. Step-by-step instructions appear on the back portion of Form No. 102 to help you complete your records. If the summary has been completed correctly, it will check out as follows: Column 1 of the second total line (Code 31) will equal the total number of cases on the OSHA Log No. 100.

POSTER REQUIREMENTS

The Occupational Safety and Health Act requires that every employer must display, in each establishment, a poster explaining the protection and obligations of employees under the Act. Where states have approved safety plans, the state poster may be displayed. Some states require that both federal and state posters be displayed. It is suggested that you consult either with the regional OSHA office or your state agency to determine whether your state requires the state or federal poster or both.

The Safety and Health Protection On the Job poster (see

job safety and health protection

The Occupational Safety and Health Act of 1970 provides job safety and health protection for workers through the promotion of safe and healthful working conditions throughout the Nation. Requirements of the Act include the following:

Employers: Each employer shall furnish to each of his employees employment and a place of employment free from recognized hazards that are causing or are likely to cause death or serious harm to his employees; and shall comply with occupational safety and health standards issued under the Act.

Employees: Each employee shall comply with all occupational safety and health standards, rules, regulations and orders issued under the Act that apply to his own actions and conduct on the job.

The Occupational Safety and Health Administration (OSHA) of the Department of Labor has the primary responsibility for administering the Act. OSHA issues occupational safety and health standards, and its Compliance Safety and Health Officers conduct jobsite inspections to ensure compliance with the Act.

Inspection: The Act requires that a representative of the employer and a representative authorized by the employees be given an opportunity to accompany the OSHA inspector for the purpose of aiding the inspection.

Where there is no authorized employee representative, the OSHA Compliance Officer must consult with a reasonable number of employees concerning safety and health conditions in the workplace.

Complaint: Employees or their representatives have the right to file a complaint with the nearest OSHA office requesting an inspection if they believe unsafe or unhealthful conditions exist in their workplace. OSHA will withhold, on request, names of employees complaining.

The Act provides that employees may not be discharged or discriminated against in any way for filing safety and health complaints or otherwise exercising their rights under the Act.

An employee who believes he has been discriminated against may file a complaint with the nearest OSHA office within 30 days of the alleged discrimination.

GPO 892-171

Citation: If upon inspection OSHA believes an employer has violated the Act, a citation alleging such violations will be issued to the employer. Each citation will specify a time period within which the alleged violation must be corrected.

The OSHA citation must be prominently displayed at or near the place of alleged violation for three days, or until it is corrected, whichever is later, to warn employees of dangers that may exist there.

Proposed Penalty: The Act provides for mandatory penalties against employers of up to $1,000 for each serious violation and for optional penalties of up to $1,000 for each nonserious violation. Penalties of up to $1,000 per day may be proposed for failure to correct violations within the proposed time period. Also, any employer who willfully or repeatedly violates the Act may be assessed penalties of up to $10,000 for each such violation.

Criminal penalties are also provided for in the Act. Any willful violation resulting in death of an employee, upon conviction, is punishable by a fine of not more than $10,000 or by imprisonment for not more that six months, or by both. Conviction of an employer after a first conviction doubles these maximum penalties.

Voluntary Activity: While providing penalties for violations, the Act also encourages efforts by labor and management, before an OSHA inspection, to reduce injuries and illnesses arising out of employment.

The Department of Labor encourages employers and employees to reduce workplace hazards voluntarily and to develop and improve safety and health programs in all workplaces and industries.

Such cooperative action would initially focus on the identification and elimination of hazards that could cause death, injury, or illness to employees and supervisors. There are many public and private organizations that can provide information and assistance in this effort, if requested.

More Information: Additional information and copies of the Act, specific OSHA safety and health standards, and other applicable regulations may be obtained from the nearest OSHA Regional Office in the following locations:

Atlanta, Georgia
Boston, Massachusetts
Chicago, Illinois
Dallas, Texas
Denver, Colorado
Kansas City, Missouri
New York, New York
Philadelphia, Pennsylvania
San Francisco, California
Seattle, Washington

Telephone numbers for these offices, and additional Area Office locations, are listed in the telephone directory under the United States Department of Labor in the United States Government listing.

Washington, D. C.
1975
OSHA 2203

John T. Dunlop

John T. Dunlop
Secretary of Labor

U. S. Department of Labor
Occupational Safety and Health Administration

illustration) may be ordered either from the U.S. Department of Labor regional or area offices or the Statistical Grant Agency in your area. A list of addresses and telephone numbers are found in the Appendix of this book. If you order by mail, provide printed, self-addressed labels with your request.

Chapter 9

Training Requirements Under OSHA

Although OSHA does not certify or approve training programs, OSHA Compliance Officers will look for evidence that an employer has provided the training required in the standards for employees.

OSHA has set up the following guidelines for training. These guidelines, when implemented, can be brought to the Compliance Officer's attention to show that the employer in "good faith" is attempting to meet standards. In the absence of those standards that are not definitive "training standards," the employer should be able to produce records indicating that his employees have received this training.

In addition, he should show the Compliance Officer that training has been given to the employee:

- Based on an analysis of the job
- Based on an analysis of the job hazards when performed by the employee

- Based on an analysis of the equipment used and practices performed by the employee
- Based on safety training priority given to the conditions and practices most likely to cause injury or illness.

The following is a list of required training standards for activities which may be used by the retailer in his operation:

CFR (Federal Register) Part	Federal Register Page	Item
1910.178	22257	Powered Industrial Trucks

Training Standard
Operator Training: Only trained and authorized operators shall be permitted to operate a powered industrial truck. Methods shall be devised to train operators in the safe operation of powered industrial trucks.

CFR Part	Federal Register Page	Item
1910.252	22305	Welding, Cutting & Brazing

Training Standard
Workmen designated to operate welding equipment shall been properly instructed and qualified to operate such equipment.

CFR Part	Federal Register Page	Item
1910.264	22331	Laundry Machinery & Operations

Training Standard
Employees shall be properly trained as to the hazards of their work and be instructed in safe practices by bulletins, printed rules, and verbal instructions.

CFR Part	Federal Register Page	Item
1910.145	22239	Safety Signs

Training Standard

All employees shall be instructed that danger signs indicate immediate danger and that special precautions are necessary. All employees shall be instructed that caution signs indicate a possible hazard against which proper precaution should be taken. Safety instruction signs shall be used where there is a need for general instruction relative to safety measures.

(Note: The item on "safety signs" reads almost as if it would be an insult to employees' intelligence. However, if you have observed how many employees will attempt to use equipment that has been marked "Defective, Do Not Use" or actually run across areas marked "Caution, Wet Floors," etc., then it is obvious why such safety sign training becomes necessary.)

CFR Part	Federal Register Page	Item
1910.151	22242	Medical Services and First Aid

Training Standard

In the absence of an infirmary, clinic, or hospital in near proximity to the workplace, which is used for treatment of all injured employees, a person or persons shall be adequately trained to render first aid. First aid supplies approved by the consulting physician shall be readily available.

(OSHA has agreed that completion of the basic American National Red Cross First Aid Course will be considered as having met the requirements. Regional OSHA administrators may determine whether other courses also are satisfactory.)

CFR Part	Federal Register Page	Item
1910.161	22247	Carbon Dioxide

Training Standard

Safety Requirements: In any use of carbon dioxide where there is a possibility that employees may be trapped in, or enter into atmospheres made hazardous by a carbon dioxide discharge, suitable safeguards shall be provided to insure prompt evacuation and to prevent entry into such atmospheres and also to provide means for prompt rescue of trapped personnel. Such safety items as personnel training, warnings signs, discharge alarms, and breathing apparatus shall be considered.

Carbon dioxide fire extinguishers and automatic systems are especially used in many retail store restaurants as well as for various other areas where CO^2 extinguishers are required.

Although there is as yet no official OSHA standard covering this (up to this writing), it is suggested that you plan to have a trained fire brigade in your store. There is a standard (1910.164) entitled Fire Brigades, but it is as yet "Reserved" pending publication to make it official. It is recommended you contact either your fire insurance company or your local fire department. Either one will gladly assist you in establishing a fire brigade training and procedure program.

Chapter 10

Fire Protection Standards

In most retail operations, the most frightening word is "Fire!" With hundreds of people working in a retail store and hundreds more people going in and out of the store everyday, is it any wonder that fire protection standards for each store as well as the OSHA regulations constitute an extremely important phase of retailing.

Here are the major reasons behind most retail store fires which cause millions of dollars in losses each year:

- Smoking and matches
- Electrical devices and wiring
- Poor housekeeping
- Flammable liquids

To control these four major fire problems:

- Have assigned smoking areas and strictly prohibit smoking anywhere else.

- Have a preventive maintenance program to check all things electrical for:
    - Equipment worn out from use
    - Improper use of approved equipment
    - Defective installation
    - Frayed wires
- Have a "good housekeeping" program whose main goal is "A Place for Everything and Everything in Its Place."
- See to it that flammable liquids are kept in proper safety cans and are correctly labeled.
- MOST IMPORTANT: Have monthly fire drills and train a store fire brigade. They can really save many a problem situation from getting out of hand.

CHOOSING THE RIGHT FIRE EXTINGUISHER

Type of Agent		Water	Carbon Dioxide	Regular Dry Chemical	ABC Dry Chemical
Class "A" Fires	Paper Wood Cloth	Excellent	Not Recommended	Not Recommended	Yes
Class "B" Fires	Gasoline Oils Paints	No	Yes	Yes	Yes
Class "C" Fires	Live Electrical Equipment, Motors, Appliances	No	Yes	Yes	Yes

MAINTENANCE OF EXTINGUISHERS

Type of Extinguisher	Maintenance
Water	Inspect monthly. Tag yearly. Hydrostatic Test every five years.

Carbon Dioxide	Inspect monthly. Weigh semi-annually. Hydrostatic Test every five years.
Regular Dry Chemical	Inspect monthly. Tag yearly. Hydrostatic Test every 12 years.
ABC Dry Chemical	Inspect monthly. Tag yearly. Hydrostatic Test every 12 years.

Fire extinguishers must be conspicuously located. It is wise to identify the pole, column, or area of fire extinguisher location by a red painted or taped band at a reasonable height and where needed, an arrow marked "fire extinguisher" to point to the location of the extinguisher. They also should be readily accessible and *immediately* available (do not block any path leading to an extinguisher).

Extinguishers should be hung no higher than five feet from the top of the extinguisher to the floor. The maximum travel distances to Class A hazard water extinguishers should be 75 feet. The maximum travel distance to Class B fires, other than for fires in flammable liquids of appreciable depth, should be 50 feet. All fire extinguishers should be tagged when inspected monthly and yearly and should have the date and name of the individual doing the inspection.

Extinguisher Rating To Meet OSHA Standards

The fire rating of an extinguisher is the guide to its extinguishing ability, rather than its size. As a result of physical testing by Underwriters Laboratories, Inc., fire extinguishers carry on their nameplates a numeral followed by a letter. These numerals and letters appear only on Class A and B types. Therefore, a fire extinguisher rated 4A can put out approximately twice as much fire as an extinguisher marked 2A.

How Many Fire Extinguishers Do You Need?

The established amount is based on recommendations that Class A type fires require 75 foot distances or one per 3,000 square feet while Class B fires require one per 50 foot distances or one per 2,500 square feet. However, the severity of the potential fire in the area, where it may occur, easy accessibility, etc. may play a very important part in making this determination.

As a retailer, you must take into consideration that children have a tendency to want to play with extinguishers in customer areas. A meeting with your local fire insurance engineer and local fire marshal to lay out the fire extinguisher locations is recommended. With the approval of the fire safety professionals, you may decide to locate the extinguishers near the stockroom entrances on the inside where customers cannot see them.

HOW FIRES START
AND ARE EXTINGUISHED

Three elements are needed to start a fire: oxygen, heat, and fuel (called the fire triangle).

Class A fires—wood, paper, textiles, and other ordinary combustibles—are extinguished by cooling and quenching with water.

Class B fires—gasoline, oil, grease, paint, etc.—are extinguished by smothering, cooling, and heat shielding.

Class C fires—in live electrical equipment—can be put out with the use of a non-conducting extinguishing agent which smothers the fire without damaging the equipment.

How To Use Fire Extinguishers

Your local fire department or fire extinguisher supplier will gladly send a representative to conduct classroom and/or field type classes.

OTHER
EXTINGUISHING SYSTEMS

Where standpipe and hose systems are provided, they should meet the design requirements of the National Fire Protection Association.

Where automatic sprinkler systems are in use, they should be maintained continuously in reliable operating condition. Periodic inspection tests should be made to ensure proper maintenance.

Chapter 11

Health Requirements

TOILET FACILITIES

Every place of employment must be provided with adequate toilet facilities, separate for each sex. The number of toilets provided should meet the following standards:

Number of Employees	Minimum Number of Toilets
1–15	1
16–35	2
36–55	3
56–80	4
81–110	5
111–150	6
over 150	One additional fixture for each additional 40 employees

Note: Where toilet facilities will not be used by women, urinals may be provided instead of toilets, except that the number of toilets in such cases shall not be reduced to less than two-thirds of the minimum specified.

When customers are permitted the use of toilet facilities on the premises, the number of such facilities should be increased appropriately according to the above table, by estimating the number of customers who will use the facilities.

TOILET AREA STANDARDS

- Toilet paper with holder must be provided for every toilet.
- Covered receptables must be kept in all toilet rooms used by women.
- One lavatory (wash basin) is required for every three required toilet seats.
- Every toilet shall have a hinged seat made of substantial material having a non-absorbent finish. Seats installed or replaced after June 4, 1973 shall be open-front type.

SEWAGE DISPOSAL

- The method used for sewage disposal shall not endanger the health of employees.

LAVATORY FACILITIES

Lavatories (wash basins) shall be made available in all places of employment in accordance with these requirements:

- In a multiple-use lavatory, 24 lineal inches of wash basin or 20 inches of a circular basin, when provided with water outlets for each space, shall be considered equivalent to one lavatory.
- Each lavatory (wash basin) shall be provided with hot and cold running water or tepid running water.
- Hand soap or similar cleansing agents shall be provided.

- Individual hand towels of cloth or paper, warm air blowers, or clean individual sections of continuous cloth toweling, convenient to the lavatories, shall be provided.
- Receptacles shall be provided for the disposal of used towels.
- Warm air blowers shall provide air at not less than 90 degrees F. and shall have means to prevent automatically the discharge of air exceeding 140 degrees F.
- Electrical components of warm air blowers shall meet electrical safety standards.

SHOWERS

Where required by a particular standard, a shower must be provided with:

- Hot and cold water feeding a common discharge line
- Soap
- Individual clean towels for drying

EATING AND DRINKING AREAS

No employee shall be allowed to consume food or beverages in a toilet room nor in any area exposed to a toxic material.

WASTE DISPOSAL CONTAINERS

Smooth, corrosion-resistant, and easily cleanable types of containers must be provided and used for the disposal of waste, food, etc. These must be emptied at least once each working day. Receptacles must be provided with a tight cover.

SANITARY STORAGE

No food or beverages shall be stored in toilet rooms.

DRINKING CUPS

A common drinking cup is prohibited. A continuing supply of single service cups should be maintained at all bottle water coolers.

MEDICAL AND FIRST AID

For the record, the OSHA standard concerning medical and first aid services and first aid personnel is as follows:

The employer shall ensure the ready availability of medical personnel for advice and consultation on matters of plant health. "Ready availability of medical personnel" means the availability of a company doctor or hospital within a reasonable distance of the retailing facilities.

In the absence of an infirmary, clinic, or hospital *in near proximity* to the workplace which is used for treatment of all injured employees, a person or persons shall be adequately trained to render first aid. First aid supplies approved by the consulting physician shall be readily available.

If the store's first-aider carries an up-to-date American Red Cross certification as a first aider or is certified by another authorized agency, this standard will be met. Of course, a Registered Nurse (R.N.) or a Licensed Practical Nurse (L.P.N.) also will meet the standard. As long as the retail establishment is open for business, a Certified First Aider or a nurse must be on the premises.

A list of first aid supplies recommended to the nurse or first aider should be kept on file in the Medical Department, and these supplies must be available at all times.

The specification, "in near proximity" to a clinic or hospital, sets no established official distance as yet, but it really should not be more than a 15 or 20 minute automobile drive away. You can check the recommendations out with the regional OSHA office.

The third and final section of the OSHA Medical and First Aid standard states, "Where the eyes or body of any person may be exposed to injurious corrosive materials, suitable facilities for quick drenching or flushing of the eyes and body shall be provided within the work area for immediate emergency use." Since many retail organizations have auto centers which use sulphuric acid for re-filling batteries, this becomes an important ruling.

In some instances local authorities have given their approval for a wash basin or sink in auto centers. They consider this adequate for "drenching and flushing" the eyes should the need arise. However, it is best to check your facility out with the regional OSHA office to determine whether this would be acceptable in your area or whether special eye wash fountains will be required.

Chapter 12

Safety Color Code for Marking Physical Hazards

The OSHA Color Code, established by the American National Standards Institute (ANSI) is intended to supplement the proper guarding or identification of hazardous conditions. As a retailer, you are required to indicate physical hazards as well as safety, firefighting, and protective equipment with identifying colors. Color should never be considered a substitute for eliminating the hazard, but its consistent use will be a benefit to employees, employer, and all others concerned. The recommended "Federal" colors are basic for their particular hazard, but within such basic color schemes there can be other color codes. For example, red is the basic color for fire, but Class A type water extinguishers are generally in silver or gold containers and have a green triangle with an "A" in the center denoting their use for Class A fires only (see section on Fire Protection Equipment Standards).

COLOR ME, RED!

Red shall be the basic color for the identification of fire protection equipment and apparatus; it is also an indication of "Danger" and a "Stop" alert.

- Fire protection equipment and apparatus applications:
  - Fire alarm boxes, fire blanket boxes, fire buckets, or pails
  - Fire exit signs, fire extinguishers, and their location backgrounds
  - Fire hose location (housing, reel and support only)
  - Fire hydrants (industrial), fire pumps, fire sirens, sprinkler piping, post indicator valves
- Danger applications:
  - Danger signs
  - Safety can or containers of flammable liquids with a flash point at or below 80 degrees F. (26.8 degrees C.) (Can or containers must be circled with a yellow band or have name of contents clearly identified in yellow.)
- Stop applications:
  - Emergency stop bars on hazardous machines
  - Emergency stop buttons or electrical switches used to stop machinery

COLOR ME, ORANGE!

Orange shall be used as the basic color for designating dangerous parts of machines or energized equipment which may cut, crush, shock, or otherwise injure, and to emphasize such hazards when enclosure doors are open or when gear, belt, or other guards around moving equipment are open or removed, exposing unguarded hazards.

Suggested applications for orange:

- Inside of moveable guards
- Safety starting buttons
- Inside of transmissions, guards for gears, pulleys, chains, etc.

- Exposed parts (edges only) of pulleys, gears, rollers, cutting devices, power saws, etc.
- Sides of conveyor belts or rollers

COLOR ME, YELLOW!

Yellow shall be the basic color for designating caution and for marking possible physical hazards such as striking against, stumbling, falling, tripping, and caught in between. Solid yellow, yellow and black stripes, and yellow and black checkers can be used interchangeably, depending on which combination will attract the most attention in the particular surroundings.

Suggested applications:

- Corner markers for storage piles
- Exposed and unguarded edges of platforms, pits, and walls
- Suspended fixtures which extend into normal operating areas
- Lower pulley blocks and cranes
- Markings for projections, doorways, travelling conveyors, low beams and pipes, frames of elevator ways, and elevator gates
- Material handling equipment
- Pillars, posts, or columns
- Striping for aisles, freight and truck loading dock edges, any runways
- Parking lots for cars

COLOR ME, GREEN!

Green shall be used as the basic color for designating "Safety" and the location of first aid equipment (other than fire fighting equipment).

Suggested applications:

- Safety bulletin boards
- Gas masks or special breathing apparatus
- First aid kits
- Entrance to first aid dispensary
- Safety deluge showers or eye wash fountains

COLOR ME, BLUE!

Blue shall be the basic color for designating caution, limited to warning against the starting, the use of, or the movement of equipment under repair or being worked on.
Suggested applications:

- Use for signs stating "Under Repair," "Out of Order," "Do Not Move," etc.

COLOR ME, PURPLE!

Purple is not a color a retailer will generally have to use; it designates radiation-type hazards.

COLOR ME, BLACK AND WHITE!

Black, white, or a combination of these two, shall be the basic colors for the designation of traffic and housekeeping markings. Solid white, solid black, single color striping, alternate stripes of black and white, or white and black checkers can be used.
Suggested applications for traffic:

- Dead ends of aisles or passageways
- Location and width of aisleways

- Stairways (risers, direction, and border limit lines)
- Directional signs

Suggested applications for housekeeping:

- Location of refuse cans
- For use underneath drinking fountains and food dispensing machines and equipment.

Chapter 13

The Office and OSHA

Who thinks of the office as an accident-producing area? Accounting, Budgeting, Customer Service, Executive Offices . . . who could possibly get hurt there? Yet as we write this, two very serious accidents which concerned an office worker and a vice-president came to our attention during the past year. The office worker fell off a rolling chair and fractured her coccyx bone, and the vice-president tripped over his own telephone wire and wound up in the hospital with a very bad back. So, you see, it can and does happen in the office, and OSHA is well aware that the office, like any other place in a company, has its potential for accidents.

The following is an office safety checklist. It includes items for which an OSHA Compliance Officer, during an inspection, can cite you if he feels that they are unsafe and could cause an accident. Make certain that you advise your office employees that when they become aware of any unsafe conditions or hazardous practices that they report these to their supervisors for correction.

OFFICE SAFETY CHECKLIST

Aisles, Floors, and Exits

- Aisles established and clear
- Holes—cracks in floor repaired
- Tripping hazards removed
- Wires not in aisles
- Entrance mats for wet weather (where entrance is directly from outdoors)
- Floors not slippery
- Carpets and rugs—secure, not rippled and with a beveled trim at edges

Stairways, Halls, and Ramps

- Handrails available and in good condition
- Stair treads in good condition
- Ramps equipped with non-slip surfaces
- Stairways not cluttered with material
- Halls clear of equipment and supplies
- Guard rails and toeboards installed on any balconies and in good condition

Office Equipment

- File cabinets secure
- File drawers kept closed when not in use
- Chairs
    - Good mechanical condition

- Springs in good condition
- Where chairs are equipped with casters, all on securely so chair can roll easily
- Fans
  - Guarded
  - Secure from falling
  - If less than seven feet from ground level, fan blade openings must be less than three-eighths of an inch or a fan blade guard must be in place
- Paper cutter blade spring functioning properly
- Paper shredders guarded
- Safe step stools or small ladders in use
- Ventilation of machines, where required
- Ammonia tanks secure and vented
- Spirit duplicating liquid properly stored in non-flammable storage facilities
- No-Smoking signs posted near spirit duplicating machines
- Paper and various other office supplies safely stacked

Bookcases, Shelves, and Cabinets

- Shelves not overloaded
- Heavy storage shelves secured to wall
- All sharp corners removed
- Safe storage on top of shelves
- Bookcases secured from tipping

Electrical Hazards

- Machines and equipment grounded
- Extension cords—three wire type
- Extension cords—maximum ten feet long

- Power cords in good condition
- Plugs and wall outlets should be in good condition (Note: Some office equipment may be used with a two-prong plug, but the tendency now is to make all office equipment with three-prongs [one for grounding] so that new office receptacles are being installed for three-prong wiring)
- Electrical switch panels clear
- Circuits not overloaded
- No wires under carpets
- All electric heaters grounded
- Electric heaters (and fans) placed so that they are not tripping hazards (nor should they be placed near flammables to create a fire hazard)

Fire Controls

- Fire extinguishers properly located and installed
- Fire extinguishers with up-to-date tags and hydrostatically pressure-tested (once every five years; once every 12 years for Halon-type)
- Fire extinguisher equipped with location arrow and not blocked
- Fire escapes clear
- Fire doors and passageways not blocked
- Approved ash trays in use
- "No-smoking" areas established as needed
- Exit lights working
- Flammable glues and liquids stored in approved metal cabinets
- Machines not overloaded
- Sprinkler heads not blocked
- Excess paper and trash removed regularly
- Fire drills held monthly

Specialized Equipment

Have employees operating special equipment been advised of any dangers concerned with the operation of the equipment—inhalation of fumes, contact with skin, ingestion? If needed, has proper protective equipment been issued, such as masks, respirators, gloves, special ventilators, etc.?

Training Program

Have office employees been safety-trained in the proper methods of:

- Safe lifting techniques
- Proper use of ladders and step stools
- Wearing of safe attire

Chapter 14

OSHA and the Auto Center

Many department stores and supermarket chains, as part of their business, have auto repair centers. And, of course, there are many retailers whose main business is auto repairs.

Retail auto repair centers are not necessarily full time garages and do not do every kind of automobile repair, but they do replace tires, do brake jobs, install batteries, etc. One might say that if it doesn't take longer than half an hour, a retail auto center will do it.

In addition to such repair centers, there are new car dealer service shops, car wash shops, and specialized auto repair businesses for mufflers or transmissions, etc.

Of course, these types of workplaces involve machinery, chemicals, and electricity. There is a need for preventive safety maintenance on various pieces of equipment. Below are the specific areas for auto centers which OSHA regulates and which must be followed.

OSHA Federal Register Section	Description of Violation
1910.157	Failure to provide an adequate number of fire extinguishers.
1910.157	Failure to mount extinguishers at proper height.
1910.157	Failure to conspicuously mark location of fire extinguishers.
1910.157	Failure to conspicuously label extinguishers with proper classification to insure that correct choice of extinguisher is put into use at time of fire.
1910.243	Failure to provide "dead man" controls on any hand-held power-driven tool which would automatically shut off the power whenever the operator releases the controls.
1910.106	Failure to store flammable liquids in approved safety containers. Also, failure to transfer such flammables into safety cans.
1910.106 (8)	Failure to store combustible waste materials in covered containers. (Such waste must be disposed of daily.)
1910.177–178	Failure to prohibit smoking in non-smoking areas.
1910.108	Failure to provide fuse-linked cover for solvent wash tank and failure to ground the portable tanks and containers used in the storage and handling of flammable and combustible liquids.
1910.309	Failure to ground plugs, power tools, and major appliances.
1910.178	Failure to have battery-charging area well ventilated, away from flammable liquids, and within 20 feet of water supply.
1910.22 B	Failure to have all walking surfaces such as aisles and service bays free from clutter and obstructions.
1910.23	Failure to provide railings and toeboards around all floor openings and/or balconies.

1910.151	Failure to provide proper first aid supplies and wash facilities.
1910.141	Failure to provide toilet room with self-closing devices and to screen toilet room entrance if visible from workroom.
1910.151	Failure to provide eye wash water facilities within the workroom area where acid is poured into batteries.
1910.37	Failure to mark exits, direction of travel where exits are not immediately visible, doors and passageways that can be confused as exits; and failure to mark other doors for their intended use.
1910.242	Failure to limit air pressure to less than 30 pounds per square inch, when used for clean-off purposes.
1910.212 ⎫ 1910.215 ⎬ 1910.243 ⎭	Failure to guard open belts, pulleys, winches, pinch points, and abrasive wheels and tool rests.
1910.252	Failure to provide check valves to prevent "flash-back" into the fuel gas system.
1910.22	Failure to mark appropriately permanent aisles and passageways.
1910.95 ⎫ 1910.132 thru 136 ⎬	Failure to provide adequate eye, ear, head, hand, foot, and respiratory protective personal equipment, where needed.

Below is a partial list of other safety items that should be complied with by auto repair centers:

- Hydraulic vehicle lifts should have their manual controls numbered to coincide with the proper lifts.
- The safety leg or hoist safety pin always must be in position before working under a hydraulic lift.
- An hydraulic lift is considered to be malfunctioning if it:

- Jerks or jumps when raised
- Slowly settles down after being raised
- Slowly rises when not in use
- Rises slowly when in use
- Comes down very slowly
- Blows oil out of exhaust line
- Leaks oil at the packing gland

If any of the above occurs, the lift should be removed from service and repaired.

- Floor drain covers should be large enough to cover their holes, and drain holes are to be cleaned periodically.

- Brake repairs and asbestos. Where brakes are repaired most of the day or where linings are machined to fit the drums, the operator should wear a dust mask to limit his exposure to the asbestos dust. Dust should be vacuumed (not blown) from the drums and the floor should be vacuumed instead of swept.

- Carbon Monoxide. Be sure there is plenty of ventilation if auto motors have to be kept running. Use a tail pipe exhaust system if, for instance, the doors have to be kept closed during the winter. Exhaust systems should be clear and hoses should not be broken. (If employees complain frequently about headaches, check carbon monoxide levels.)

- Hydrogen. The battery-charging area produces hydrogen gas. You must provide adequate ventilation and post "NO SMOKING" signs for this area.

- Jacks should be inspected once every six months, and all cars on jacks should be blocked or secured at once.

- Welding, cutting and brazing equipment
  - All cylinder valves must be closed when not in use
  - Acetyline is not to be at a pressure in excess of 15 psi gauge or 30 absolute
  - All cylinders must be secured with a chain-type lashing
  - Proper eye protection for welding, cutting, or braz-

ing must be made available and kept in good repair and sanitary condition

- All storage lofts, second floors, etc. must display floor-load capacity signs.

- All ladders must be in good condition and meet prescribed OSHA standards. Check their treads for any broken rungs.

- Battery-filling operations (using sulphuric acid) must provide rubber goggles, rubber gloves, and aprons near at hand.

- Concerning hand tools: Replace mushroom-head chisels, broken hammer handles, worn or bent wrenches. They are a constant source of accidents.

- All electrical circuit breakers should be identified to easily ascertain the function they perform.

- Make certain that an OSHA Employee Poster is displayed prominently and that all injuries are reported to the properly designated medical personnel for treatment and recordkeeping, according to OSHA regulations.

Chapter 15

Warehousing and OSHA

RETAILERS' WAREHOUSES

Many retailers, as part of their businesses, maintain warehouses for merchandise. A small retailer's "warehouse" may be just a few stockrooms in back of his store, but these are quite important to the success of his business. Some major department stores have warehouses that are three or four football fields long and keep merchandise stacked 28 or more feet high.

One of the major activities in warehousing is materials handling. Materials handling accidents in all industries account for 25 percent of all the lost time injuries suffered in the United States. Needless to say, if we can make even a ten percent improvement in cutting down on materials handling accidents, it would bring a substantial gain in manpower on the job and lowering of accident expenses. This is where OSHA standards play an important role.

Management is aware that periodic plant inspections followed by immediate corrective action of unsafe conditions and hazardous practices can generally reduce accidents. The following pages which cover OSHA standards pertaining to warehousing operations can serve as a guideline to the warehouse manager in making a walk-through inspection of the plant. The standards are identified by section numbers as they appear in the Federal Register.

OSHA
Standards Pertaining
To Warehousing Operations

Walking/Working Surfaces

1910.23 *Hatchway and Chute Floor Openings*:
Hinged floor opening cover, closed when not in use or removable guard railing and toeboards.

1910.23 *Stairway Railing and Guards*:
Stairs having four or more risers:
a) Less than 44 inches wide both sides enclosed, at least one handrail on right side descending.

b) Less than 44 inches wide both sides open, one railing on each side.

c) More than 44 inches, but less than 88 inches wide—one railing on each side.

d) More than 88 inches wide—one railing on each side and one intermediate railing. Stair railing vertical height not more than 30 inches to tread.

1910.23 *Platforms*:
Every open-sided floor or platform four feet or more above adjacent floor or ground level shall be guarded by a standard railing with a standard toeboard.

1910.25 *Portable Ladders*:
a) Maintained in good condition at all times. Inspected frequently, defective ladders tagged for repair or destruction, "Dangerous—Do Not Use".

b) Provide safety feet.

c) Metal bearings of locks, wheels, pulleys, etc. shall be lubricated frequently.

d) Store properly.

e) Don't use metal ladders when doing electrical work.

1910.27   *Fixed Ladders:*

a) Minimum diameter of three-quarters of an inch for metal ladders, and one and one-eighth inches for wood ladders (rungs and cleats).

b) Distance between rungs, cleats and steps shall not exceed 12 inches and be uniform.

c) Minimum clear length of rungs or cleats shall be 16 inches.

d) Ladders over 20 feet in height shall be provided with cages.

1910.27   *Water Tower Ladders:*

Either caged or provided with lifebelts, friction brakes, and sliding attachments.

1910.35   *Means of Egress:*

a) At least two means of egress, remote from each other.

b) Unobstructed passageway to exits.

c) Exits visible or route to reach exit conspicuously indicated.

d) No blocking of fire exits.

e) Exit doors shall swing out with exit travel.

f) Devices or alarms on doors designed to restrict improper use of doors cannot prevent emergency use of exits.

g) No flammable decorations or furnishings shall be used.

h) Exits shall be marked with an "EXIT" sign not less than 6 inches high, suitably illuminated with not less than 5 foot-candles of light.

i) Any door, passage or stairway not an exit, or likely to be mistaken for an exit shall be idenfied "Not an Exit" or "To Basement", "Storeroom", etc.

Storage of Hazardous Materials

1910.106   *Flammable Liquids*:

Storage Cabinets:

Metal—Bottom, top, door and sides at least No. 18 gauge sheet iron, double walled with one and one-half inches air space. Door provided with a three-point lock, door sill raised at least two inches above bottom of cabinet.

Wood—Bottom, sides and top constructed of at least one inch plywood, hinges mounted so as not to lose holding capacity due to burning out of screws.

Not more than 60 gallons of flammable (flashpoint below 140 degrees F.) or 120 gallons of combustibles (flashpoint above 140 degrees).

Permissible Amounts of Flammable or Combustible Liquids:

Offices—Prohibited except that which is required for maintenance and operation of equipment.

Mercantile—(Sales Areas) Shall not exceed two gallons per square foot of gross floor area of that portion of the store actually being used for merchandising flammable and combustible liquids.

(Stock Areas) Where the aggregate quantity of additional stock exceeds 60 gallons of Class IA, or 120 gallons of Class IB, or 180 gallons of Class IC, or 240 gallons of Class II, or 500 gallons of Class III, or any combination of Class I or Class II exceeding 240 gallons, it shall be stored in a room that complies with construction requirements.

		Protected Storage Maximum per pile		Unprotected Storage Maximum per pile	
		Gallons	Height	Gallons	Height
Class IA	Flashpoint below 73°F—Boiling point below 100°F	2,750	3 ft.	660	3 ft.
Class IB	Flashpoint below 73°F—Boiling point above 100°F	5,500	6 ft.	1,350	3 ft.
Class IC	Flashpoint above 73°F—Boiling point below 100°F	16,500	6 ft.	4,125	3 ft.
Class II	Flashpoint above 100°F	16,500	9 ft.	4,125	9 ft.
Class III	Flashpoint above 140°F	55,000	15 ft.	13,750	12 ft.

Personal Protective Equipment

1910.135 *Head Protection*
1910.136 *Foot Protection*

General Environmental Controls

1910.141 *Housekeeping*:
All places of employment shall be kept clean and orderly.

Color Coding
Red   Fire Protection Equipment:
Fire Buckets
Fire exit signs
Fire extinguishers
Fire hose locations (on reel, supports and housing)
Fire hydrants
Fire pumps
Fire sirens
Post indicator sprinkler valves
Sprinkler piping

Danger—Safety cans or other portable flammable liquid containers (red with yellow band around can or name of contents stenciled or painted in yellow).

Machines—Stopping devices shall be red.

Orange   Dangerous and/or moving machine parts.

* *Yellow*   Marking physical hazards, i.e., low overheads, tripping hazards, etc. (* Yellow or yellow and contrasting color).

Green   Safety equipment or first aid equipment.

Blue   Caution against starting equipment being repaired or worked on.

Purple   Radiation Hazard.

Black, White, or combinations   Traffic and housekeeping markings.

1910.151   *Medical and First Aid*:

a)   Ready availability of medical personnel for advice and consultation for plant health.

b)   In absence of infirmary, clinic or hospital in near proximity, trained First Aider(s). First aid supplies shall be readily available.

Fire Protection

1910.157   *Portable Fire Extinguishers*:

Class A   Wood, cloth, paper

Class B   Flammable liquids, gases, and greases

Class C   Electrical equipment

Class D   Combustible metals, magnesium, titanium, zirconium, ABC (All purpose extinguishers).

Fire extinguishers shall be maintained in fully charged condition, kept in designated places at all times. Shall not be obstructed or blocked from view. If extinguishers intended for different classes of fires are grouped, their intended use shall be marked conspicuously.

Mounting of Extinguishers—Gross weight not exceeding 40 pounds if top of extinguisher not more than five feet above floor; over 40 pounds not more than three and one-half feet from floor.

Light hazard   Offices, schoolrooms, churches.

Ordinary hazard   Auto showrooms, parking garages, light manufacturing warehouses.

Extra hazard   Severe fires, woodworking, auto repair, high piled warehouses (over 14 feet).

Types of extinguishers required for extra hazard (piles over 14 feet):

Class A   Loaded stream, multi-purpose dry chemical, water types.

Class B   Bromotrifluoromethane, carbon dioxide, dry chemical, loaded stream, multipurpose dry chemical.

Class A Hazards

Ext. Rating	Max. travel Distance to Extinguishers	Sq. Ft. Light Hazard	Sq. Ft. Ordinary Hazard	Sq. Ft. Extra Hazard
1A	75 Feet	3,000	—	—
2A	75 Feet	6,000	3,000	—
3A	75 Feet	9,000	4,500	3,000
4A	75 Feet	11,250	6,000	4,000
6A	75 Feet	11,250	9,000	6,000

Class B Hazards

Hazard	Minimum Ext. Rating	Maximum Travel Distance
Light	4B	50 Feet
Ordinary	8B	50 Feet
Extra	12B	50 Feet

Inspection of Extinguishers:

Shall be at least monthly, to ensure they are in designated place and operable. Examined and/or recharged annually. Tagged with date and who performed recharge.

Hydrostatic Testing:

Soda acid, cartridge water, foam, loaded stream, dry chemical with stainless steel, aluminum or soldered brass shells, carbon dioxide *every 5 years*. Dry steelshells, bromotrifluoromethane every 12 years.

Record tag of metal with date of test, test pressure, agency making test.

Sprinkler Systems:

Shall have at least one automatic water supply of adequate pressure, capacity and reliability.

Fire Department Connection:

One or more connections shall be provided in all cases. Pipe size not less than four inches. No shut-off valve in Fire Department connection. Hose connections shall be designated by sign—raised letters at least one inch in size.

Sprinkler Alarms:

Waterflow alarms shall be provided on all sprinkler installations.

Electric Signaling Systems:

Sprinkler systems should be supervised by a central station system.

Storage Types:

I. Over 15 feet but not more than 21 feet in solid piles, or over 12 feet but not more than 21 feet in piles containing horizontal channels.

II. Not over 15 feet high in solid piles or not over 12 feet in piles containing horizontal channels.

III. Commodities incapable of producing a fire that would cause appreciable damage.

Sprinkler Head Clearance

Type I   36 inches

Type II   18 inches

Type III   18 inches

Fire Brigades:

Develop and institute plant emergency organization program for fire fighting and evacuation purposes.

Materials Handling and Storage:

- Storage of materials shall be stable and secure against collapse.

- Clearance limit signs shall be provided to warn of limits.

- Derail and/or bumper blocks shall be provided on railroad tracks.

1910.178   ### Powered Industrial Trucks:

All name plates and markings should be in place and maintained in legible condition.

Approved trucks for storage warehouses (nonhazardous or small amount of combustibles):

D.—Deisel engine powered

E.—Electric

G.—Gasoline

LP.—Liquified Petroleum

Safety Guards: High lift rider trucks shall be provided with an overhead guard unless operating conditions do not permit. Load backrest extension shall be used when necessary to minimize the possibility of the load or part of it from falling backward.

Battery Charging: Facilities shall be provided for flushing and neutralizing spilled electrolyte, for fire protection, for protecting charging apparatus from damage to trucks, and for adequate ventilation.

When racks are used for support of batteries, they

should be made of nonconductive materials or be coated or covered with a non-conductive material. A conveyor, hoist or equivalent equipment shall be provided for handling batteries.

A carboy tilter or siphon shall be provided for handling electrolyte.

When charging batteries, acid shall be poured into water, water shall not be poured into acid.

When charging batteries, the vent caps should be kept in place to avoid electrolyte spray. The battery (or compartment) cover shall be open to dissipate heat.

Smoking shall be prohibited in the charging area.

Operator Training (Also see suggested operator training program at end of chapter):

Only trained and authorized operators shall be permitted to operate a powered industrial truck. Methods shall be devised to train operators in the safe operation of powered industrial trucks.

A powered industrial truck is unattended when the operator is 25 feet or more away from the vehicle, which remains in his view, or whenever the operator leaves the vehicle and it is not in his view. When the operator is within 25 feet and the truck is in view, the load engaging means shall be fully lowered, controls neutralized, and the brakes set.

Wheel chocks shall be in place to prevent movement of trucks, trailers or railroad cars while loading or unloading.

When lifting personnel, a safety platform shall be used with means provided whereby personnel on the platform can shut off power to the truck.

Fire aisles, access to stairways and fire equipment shall be kept clear.

Industrial trucks shall be examined before being

placed in service on a daily basis, or after each shift if used on an around-the-clock basis.

Dockboards or bridgeplates shall be properly secured before they are driven over.

1910.12 *Machine Guards*:
Fans—
Less than seven feet from the floor shall be guarded. The guard shall have openings no larger than three-eighths inch.

Circular saws—
Provided with an automatically adjustable hood guard, spreader, and non-kickable fingers.

Grinding wheels—
Work rests adjusted to one-eighth inch. Periphery guarding shall not exceed one-quarter inch.

1910.241 *Hand and Portable Power Tools*:
Compressed air shall not be used for cleaning purposes except where reduced to less than 30 P.S.I.

Portable circular saws having a blade diameter greater than 2 inches shall be equipped with guards above and below the base plate, and have a constant pressure switch.

All portable electric tools shall be properly grounded.

The following offenses have been cited most frequently by Compliance Officers during inspections of retail and supermarket warehouses:

**Federal Register
Section**

1910.37 (Q) (1)	Unmarked exit
1910.157 (a) (3)	Obstructed fire extinguisher

1910.22 (a) (1)	Extension cords, telephone wires (tripping hazard)
1910.157 (E) (1) (i)	Extinguisher mounted in excess of five feet
1910.215 (a) (4)	Bench grinder workrest not adjusted within one-eighth inch
1910.215 (b) (9)	Bench grinder with no upper peripheral guard
1910.157 (d) (3) (i)	Fire extinguisher not inspected
1910.37 (Q) (8)	Exit markings only one and one-half inches high
1910.178 (G) (10)	Smoking not prohibited in battery charging area
1910.157 (a) (1)	Extinguisher on fork truck discharged
1910.176 (a)	Pallets extending out into aisleway
1910.178 (N) (11)	Use of flat, unsecured dock plate
1910.23 (c) (1)	No handrail or toeboard on overhead platform
1910.22 (b) (2)	Failure to properly mark aisles and passageways
1910.157 (d) (4) (iii)	Failure to test fire extinguishers hydrostatically
1910.178 (k) (1)	Failure to place wheel chocks under trucks
1910.132 (a)	Failure to provide hard hats for warehouse employees
1910.178 (m) (3)	Employee riding on box being transported by fork truck
1910.309 (a)	Ungrounded refrigerator
1910.309 (a)	Unmarked feeder or branch circuit boxes
1910.157 (c) (2) (i)	Insufficient number of fire extinguishers
1910.159 (E) (2)	Failure to maintain 18-inch clearance of stock from sprinklers
1910.22 (D) (1)	Floor loading not posted
1910.176 (B)	Leaning stack of storage materials
1910.212 (a) (3) (ii)	Unguarded box stapler
1910.178 (m) (13)	Removed spinner knobs from fork trucks

SUGGESTED POWERED
INDUSTRIAL TRUCK OPERATORS
TRAINING PROGRAM

The heart of any warehousing operation is the safe movement and storage of material. Powered industrial trucks and their intelligent and safe operation represent the vitality of your materials handling program. Therefore, training must play an important part in a successful operation.

OSHA regulations state, "Only trained and authorized operators shall be permitted to operate a powered industrial truck." Unfortunately, the only requirement that OSHA adds is, "Methods shall be devised to train operators in the safe operation of powered industrial trucks." OSHA does not spell out any standards as to what the training "methods" should be.

Therefore, we are going to present an idea of what a powered industrial truck operator training program could include as a guide to developing your own training program. Naturally, your program will fit your own particular needs, but it should touch on these items. The do's and don'ts of these safety rules are based on the experience of others; sometimes they were learned the hard way—"by accident."

Selection of Operators

In the selection of qualified powered industrial truck drivers, the following items are important:

Physical Qualifications
- Good physical and mental condition with no diseases or disabilities that could interfere with the safe and efficient use of the industrial truck

- Normal reflexes and reaction time

- Binocular vision

- Vision in each eye no less than 20/40 corrected

- Corrective safety eyeglasses must be worn if needed

- Depth perception no less than 90 percent of normal

- Normal hearing, preferably without the use of a hearing aid

Other Qualifications
- Minimum age of 18 for any industrial truck driver
- Able to read posted signs

Classroom Instructions

- All powered industrial truck operators must attend a "General Safety Rules" indoctrination geared specifically to operation of powered industrial trucks for a period of not less than one hour.
- The operator must achieve a "pass" mark when taking a written test based on his knowledge of the "General Safety Rules."

GENERAL SAFETY RULES

1. Load Capacity
   - Know the load capacity of vehicle. Do not attempt to handle loads exceeding rated capacity.
2. Vision
   - Loads should not obstruct vision. If vision is obstructed, drive vehicle in reverse.
3. Hands
   - Never drive with wet or greasy hands.
4. Feet and Legs
   - Keep feet and legs inside the running lines of vehicle.
5. Starting
   - Before starting, always check out your vehicle for:
     - Bad brakes
     - Bad wheels
     - Defective steering
     - Lights
     - Horn defect
     - Any defect in the mechanism

Report all unsafe conditions to a supervisor. Do not attempt to repair vehicle yourself.

6. Traveling

   - When tiering or stacking, avoid sudden stops. (This tends to shift the load.)

   - Use extreme caution when handling loads that extend above the top of the mast. If truck is equipped with a tilt device, use it.

   - Tilting the load toward you greatly improves stability.

   - Slow down when approaching any danger points such as intersections, aisles, obstructions, inclines, or congested areas.

   - Center the load on forks when picking up a pallet or skid; uneven or offside loading can cause load to shift unexpectedly.

   - Observe all traffic regulations. Keep to the right. Try to avoid traveling in congested areas during breaks or shift changes.

   - Slow down for wet or slippery floors. Report any unsafe floor conditions. Hitting a piece of wood or debris can cause the wheel to spin suddenly.

   - Do not carry passengers or allow any unauthorized persons to operate your vehicle. Always face or look in direction of travel.

7. Parking

   - Whenever the vehicle is left unattended, the parking brake should be applied, forks placed flat on floor, and key removed from ignition.

8. Make use of all convex mirrors at all hidden corners.

9. Use your horn intelligently. Signal at intersections, corners, doorways, or when approaching a pedestrian.

   - REMEMBER—the pedestrian has the right of way.

10. Look before you change directions and when lowering the load. Never allow anyone under a load or under the forks or carriage . . . *loaded or empty!*

11. Do not allow "horseplay" around your truck.

12. Never carry flammables or acids unless they are in approved safety containers.

13. Do not drive in the dark. Either be sure your fork truck has adequate lighting or the area has been adequately illuminated.

14. If you have an accident, stop immediately and investigate. Report all accidents to your supervisor immediately.

Upon completion and passing of the written portion of the training test, the employee must then pass your test for the actual operation of the vehicle. When he has proven that he operates the vehicle properly, a company license stating that he is an authorized and trained operator of that vehicle should be presented. Where he is to operate other industrial-powered vehicles, actual training and driving tests should be given for each vehicle, and the license should indicate each vehicle for which he has been trained and tested. Be sure to explain that the license issued is only good while the person is in your employ and that it is not transferable to other companies.

Chapter 16

Suggested Safety Programs and Procedures

The following are some suggested safety programs that will help the retailer meet the "good faith" requirements of OSHA.

THE NEW EMPLOYEE

- Arrange for a pre-employment physical (either a doctor's examination or have the prospective employee fill out a health questionnaire).
- During the training of the new employee, by either your training department or his immediate supervisor, these five safety questions and answers should be discussed.

  Q. What do I do if I get hurt?

  A. Report the injury to your supervisor and go to the medical department (or the assigned first aider) so that it can be treated. Do not attempt to treat it yourself or disregard it.

Q. What do I do if I see an unsafe condition or a hazardous practice?

A. Report the unsafe condition or hazardous practice to your supervisor. If at all possible, write it to him rather than just telling him. He should have it removed, repaired or replaced. If there is a member of the Safety Committee around, report the unsafe condition or hazardous practice to him as well.

Q. What do I do in the event of an emergency? That is, if I see a fire or a pipe broken and water gushing out?

A. If you become aware of an emergency situation:

   a) Keep calm—don't panic

   b) Go to the nearest telephone and calmly tell the operator of the emergency. Be sure you give her the best approximate location, then proceed to follow the same exit procedures you practice in store fire drills. (Of course, the employee should be made aware of just what these procedures are.)

Q. How do I stop an escalator?

A. At each escalator, a few inches from ground level, you will see a "stop" button. Before you press it, be sure to yell "HOLD TIGHT, PLEASE," then press the button to stop the escalator. Sometimes these buttons are covered by small awning-type covers to prevent children from playing with them. For this type, just reach in and press the button. Some of the very latest models have red stop buttons on the balustrade sides, and some even have the buttons not on the escalators but rather on columns close by. Check with your supervisor to be sure of their locations. You cannot start an escalator. You must call the maintenance department and they will start it with a key after first examining the reason for its being shut down. Be sure to learn the exact locations of the buttons for the escalators in your store.

Q. Who is responsible for my safety in the company?

A. You are! And your supervisor and everyone who works in the company, from the president to the porter. Anyone who commits an unsafe act has no guarantee as to who will or won't get hurt. Remember, through an unsafe act, whether your own or someone else's, you can become the injured party. So obey all the company's safety rules, wear protective equipment, don't smoke in non-smoking areas, and don't take unnecessary chances.

New Employee Safety Booklet

● Make up a safety rules and regulations booklet to be issued to each employee.

SAFETY COMMITTEE

Set up a formal Safety Committee program with monthly reported meetings to review action taken on recommendations made by the committee members on unsafe conditions and hazardous practices. A formal Safety Committee program should consist of the following:

● Identification badge for each committee member.

● Required attendance at all committee meetings (monthly).

● Designating a store safety coordinator to act as the formal Safety Committee Chairman, responsible for setting up meetings and following up safety recommendations.

● Assigning each member to a certain area of the store, to make a formal safety inspection of that area on a monthly basis. Findings should be reported, in writing, and submitted to the Safety Committee Chairman.

● The Safety Committee Chairman should review the recommendations made at each meeting and follow up on those "in progress" or "not yet completed." It is suggested that the safety chairman/coordinator design an easy form for committee members to use to make recommendations.

- Minutes should be kept for each meeting, and copies of the minutes should be sent to all involved store executive personnel as well as to members of the Safety Committee.

- Minutes should be written up by the store safety coordinator or his designee.

- It is suggested that the store safety coordinator work up a formal Safety Committee manual outlining, in detail, the duties of store Safety Committee personnel. This is especially valuable since there is generally a high turnover in such personnel due to transfers, promotions, etc.

Suggested Agenda for a Safety Committee Meeting

- Review the employee injuries for the past month. (Also review any customer accidents.)
- Review recommendations from prior meetings.
- Discuss this month's inspections.
- Plan a special safety project for the coming month for review. Subjects might be falls, strains, ladders, housekeeping, or fire extinguishers.
- Open discussion on any safety subject.
- Set a date for the next meeting and adjourn.
- (Be sure to keep the committee minutes up-to-date. The OSHA Compliance Officer usually will ask to see them.)

FIRE DRILLS

- Arrange for monthly fire drills

- Your local fire department will gladly work with you to plan a monthly fire drill procedure as well as train a store fire brigade.

KEEP SAFETY ALIVE THROUGH:

- Contests
- Safety promotions
- Newsletters and news bulletins
- Posters
- Off-the-job safety programs

KEEP OSHA AND OTHER RECORDS UP-TO-DATE

- Keep your OSHA Forms 100 and 102 up-to-date
- Keep your employee and customer accident reports up-to-date and show how you investigated them. (See illustration of accident investigation report on page 115.)

WORK CLOSELY WITH YOUR INSURANCE CARRIER

Use all the services and technical know-how made available to you.

INSPECTIONS

For multi-store operations, make a formal weekly inspection—a different store each week on a year-round basis. Arrange to have your insurance company safety engineer accompany you on each of these trips. Here is a suggested checklist for a department store inspection. It can easily be adjusted with a few changes to become a working supermarket inspection checklist.

LOSS CONTROL AND ENGINEERING DEPARTMENT

REPORT OF ACCIDENT INVESTIGATION

Name of Company: ....

Address of Plant: .... Plant No.: ....Time Charge:.................

1. Name of injured _____ 2. Age _____

3. Department_____ 4. Occupation _____

5. Date of accident _____ 6. Time____A.M. 7. Hours worked this shift prior to accident_____
                                              P.M.

8. Accident location _____

9. NATURE OF INJURY

Abrasion ☐	Dermatitis ☐	Fracture ☐	Puncture ☐
Amputation ☐	Dislocation ☐	Infection ☐	Sprain, Strain ☐
Contusion ☐	Foreign body ☐	Laceration ☐	Other _____

10. PART OF BODY INJURED

Eyes ☐	Back ☐	Fingers ☐	Toes ☐
Head, Neck ☐	Arms ☐	Legs ☐	
Body ☐	Hands ☐	Feet ☐	Other _____

11. ACCIDENT AGENCY — That which was most closely associated with the injury:

Animals, Insects, etc. ☐	Electrical apparatus ☐	Radiation ☐
Boilers, Pressure Vessels ☐	Elevators, Hoists, Conveyors ☐	Vehicles ☐
Chemicals (fumes, vapors) ☐	Hand tools ☐	Working surfaces ☐
Dusts (asbestos, silica, etc.) ☐	Machines, Pumps, Engines ☐	Other ☐

12. ACCIDENT TYPE — How injured:

Caught in, on or between ☐	Inhalation, Absorption, ☐	Over-exposure: radioactivity ☐
Electrical contact ☐	Ingestion ☐	temperature ☐
Falls: to same level ☐	Over-exertion: strain, sprain ☐	Striking against, Stepping on ☐
to different level ☐	hernia	Struck by moving or flying objects ☐

13. UNSAFE CONDITION: Physical cause of accident:

Defective conditions ☐	Improper ventilation ☐
Hazard arrangement or procedures ☐	Unsafe design or construction ☐
Improperly guarded ☐	Unsafe dress or apparel ☐
Improper illumination ☐	No unsafe condition ☐

14. UNSAFE ACT: Personal cause of accident:

Distracting, Teasing, Abusing ☐	Taking unsafe position or posture ☐
Failure to use safe attire ☐	Using unsafe equipment ☐
Making safety devices inoperative ☐	Unsafe loading, mixing, placing ☐
Operating without authority ☐	Working on moving or dangerous equipment ☐
Operating or working at unsafe speed ☐	No unsafe act ☐

15. UNSAFE PERSONAL FACTOR — Accident subcauses:

Bodily defects ☐	Lack of knowledge or skill ☐	Other _____
Improper attitude ☐	No unsafe personal factor ☐	

16. Brief description of accident: _____

_____

_____

17. What action has been taken to prevent similar accidents?_____

_____

_____

Signed By: _____ Title: _____ Date:_____

AUTHORS' PERSONAL
OSHA CHECKLIST

1. OSHA poster is posted in appropriate location.

2. OSHA Form 100 (Record of Occupational Injury and Illness) maintained.

3. Regular employee and customer accident reports, with any follow-up reports are kept in orderly fashion.

4. All new employees are being indoctrinated in safety by films, pamphlets, etc.

5. Parking lot meets requirements of striping, speed limit signs posted, lighting maintained, snow and ice removal arrangements, hole repairs and debris cleaning and removal.

6. Stairways clear of any debris or obstructions.

7. Exits are marked properly. Exit lights not hidden.

8. Exits must not be blocked and must be clearly marked.

9. All floor mats must not slide or be curled up and must be butted.

10. Glass panels marked by decals.

11. All stairways more than 88 inches wide have a handrail down the middle.

12. Main aisles must be at least 36 inches wide; keep aisles clear of boxes, displays, wires, etc.

13. Fixtures should not have extensions which can trip people, or sharp corners.

14. All display counters should be free of nails, broken glass or easily knocked over counter displays.

15. Stock drawers must be closed when not in use.

16. Check for open carpet seams, tears or folds in carpet.

17. Check for holes in floor, floor tiles missing, slippery floor finishes.

18. There are no obstructions in or near elevators. Escalators are maintained properly.

19. Doors that are not exits must be marked "not an exit," "office," "basement," etc.

20. Any door which is an exit must be unlocked when employees are in the building.

21. All aisles must be properly striped; no stock, empty cartons or debris piled in aisles.

22. Material or stock is not stacked within 18 inches of sprinkler head.

23. Hazardous and flammable material stored properly.

24. Employees trained in use of power equipment; forklift operators properly trained.

25. All rolling ladders have wheel locks and rubber "crutch" tips.

26. All pulleys are guarded; all saw blades have guards.

27. All grinders have hood guards; "table" not more than one-eighth inch away from grinder.

28. All equipment designed with interlocks should not have interlocks made inoperable.

29. All electrical equipment, wall sockets and plugs—three-prong types and grounded.

30. Extension cords not to be strung across walk areas without being properly covered.

31. All electrical hand tools should be grounded.

32. "Temporary wiring installations" are not acceptable.

33. Fan blades less than seven feet off the floor must be "mesh guarded."

34. Fire extinguishers readily accessible, charged within past year, hung properly.

35. Fire doors not blocked and work properly.

36. No smoking signs posted where necessary; no smoking rules enforced.

37. Monthly fire drills.

38. Emergency lighting system works properly.

39. Floors are clean and free of grease and spills.

40. Flues, ducts and hoods are clean, relatively free of grease.

41. Meat cutting and all other machines have guards on blades, pulleys and cutting edges and are grounded.

42. Refrigerators and other food and storage areas are kept clean and at proper temperatures.

43. No excessive noise levels (over 90 decibels).

In addition, following is a list of "30 Most Frequently Violated Standards" for which OSHA Compliance Officers have been citing retailers. Following that list are two safety checklists which retailers have found valuable in maintaining safe conditions.

30 MOST FREQUENTLY VIOLATED OSHA STANDARDS

1. 1910.309 (a)   National Electrical Code
   (I Grounding)
   (II Workmanship)
   (III Marking controls)
2. 1910.219 (d) 1   Pulley guarding
3. 1910.215 (a) 4   Work rests for grinders
4. 1910.252 (a) 2   Oxygen and flammable gas cylinder storage; (welding)
5. 1910.212 (a) 1   Machine guarding, general; type of guarding
6. 1910.212 (a) 5   Exposure of fan blades
7. 1910.23 (c) 1   Protection of open sided floors and platforms
8. 1910.22 (a) 1   Housekeeping, general
9. 1910.037 (q) 1   Exit marking signs—"no exit" signs
10. 1910.219 (e) 1   Guarding belt, rope, and chain drives
11. 1910.215 (a) 2   Guard design for grinders
12. 1910.157 (d) 3   Yearly service and maintenance of fire extinguisher
13. 1910.242 (b)   Compressed air used for cleaning (30 PSI max)

14.	1910.212 (a) 3	Point of operation guarding
15.	1910.215 (b) 9	Guarding of abrasive wheel machinery (Tongue guard)
16.	1910.213 (h) 1	Radial saw guards and sign
17.	1910.219 (f) 3	Guarding of sprockets and chains
18.	1910.22 (b) 2	Aisles and passage ways—marking
19.	1910.157 (a) 5	Mounting of fire extinguishers
20.	1910.22 (d) 1	Floor load limits marked
21.	1910.157 (a) 2	Location of fire extinguishers, and marking of location
22.	1910.23 (d) 1	Stairway railing and guards
23.	1910.106 (e) 2	Incidental storage of flammable liquids
24.	1910.25 (d) 1	Care and use of portable wood ladders
25.	1910.213 (h) 1	Hoods for swing saws
26.	1910.176 (b)	Stacking and piling of material in storage
27.	1910.151 (c)	Emergency eye wash and drench showers
28.	1910.157 (d) 2	Monthly inspection of fire extinguishers
29.	1910.132 (a)	Personal protective equipment
30.	1910.151 (b)	First aid kit and first aid certification

OSHA PROGRAM CHECKLIST

I   **Statute and Standards**

    (A)   Is a copy of the standard (June 27, 1974 issue of the Federal Register) readily available?   YES__ NO__

    (B)   Is a program in effect to bring the location into compliance with the standards?   YES__ NO__

II   **Recordkeeping**

    (A)   Are forms 100, 101, or equivalent and 102, properly completed.   YES__ NO__

    (B)   Are forms 100, 101 and 102 filed properly so that they can be maintained for a minimum of five years?   YES__ NO__

III *Posting Notice*
    (A)   Is the official posting notice properly posted in a conspicuous place?   YES__ NO__

IV *Medical Services*
    (A)   Does the location have a medical dispensary readily available in case of accident or illness?   YES__ NO__

    (B)   If a hospital or clinic is not nearby, is a qualified first-aider on hand during all dispensary off hours?   YES__ NO__

    (C)   Are first aid supplies available for use during dispensary off hours?   YES__ NO__

    (D)   Have the first aid supplies been approved by a physician?   YES__ NO__

V *Employee Training*
    (A)   Have employees been made aware of the law and their responsibilities under it? How?   YES__ NO__

    (B)   Have employees been properly trained in compliance with the training requirements of the OSHA standards? (e.g. powered industrial trucks, power presses, chemical handling, etc.)   YES__ NO__

VI *Supervisory Training*
    (A)   Have meetings been held with all supervisors explaining the law and their responsibilities under it?   YES__ NO__

    (B)   Are supervisors properly trained to recognize unsafe conditions or unsafe practices?   YES__ NO__

VII *Inspections*
    (A)   *By Supervisors*
        1.   Are supervisors making continuous inspections of their areas of jurisdiction?   YES__ NO__

        2.   Do you feel supervision is responsive to employee complaints that can be cause for an OSHA inspection?   YES__ NO__

3. Do supervisors satisfactorily follow-up on reported unsafe conditions until corrected?   YES__ NO__

(B) *By Safety Committees*
1. Does your location have inspection committees?   YES__ NO__

2. Is a formal inspection report prepared?   YES__ NO__

3. Are members of the inspection committees trained in what to look for and how?   YES__ NO__

VIII *Safety Rules*
1. Does the location have a set of general safety rules in effect?   YES__ NO__

2. Have more specific safety rules covering hazardous equipment, materials, chemicals, etc., been developed?   YES__ NO__

3. In your opinion, are the general and specific safety rules adequately enforced?   YES__ NO__

IX *Preparation For OSHA Visit*
(A) Has management been advised concerning what to expect when the compliance officer visits the location?   YES__ NO__

(B) Has a company representative(s) been designated to accompany the compliance officer on the tour of inspection?   YES__ NO__

X *Citations*
(A) In the event a citation is received, have arrangements been made to review each violation and extent of penalty with the legal, plant engineering and industrial relations functions?   YES__ NO__

(B) Has a procedure been established for the timely contesting of citations and/or penalties, if warranted?   YES__ NO__

(C) Has the store been advised as to the need for *prompt* notification to division staff concerning the commencement of the inspection, findings and other relevant details?   YES__ NO__

(D)   Has the store established a procedure for prompt referral of the formal citation and penalties to division for timely review?   YES__ NO__

(E)   Has the store established a workable relationship with the OSHA area office for discussion of questions concerning the standards, citations, penalties, etc.?   YES__ NO__

INSPECTION CHECK LIST

Fire Prevention

	YES	NO
Are smoking areas approved?	( )	( )
Are "No-Smoking" signs posted in unauthorized areas?	( )	( )
Are flammable liquids in safety containers?	( )	( )
Are containers holding substances that are a hazard properly labeled?	( )	( )
Are flammable liquids limited to daily supply?	( )	( )
Do drums contain self-closing draw-off faucets?	( )	( )
Do outside storage tanks containing flammable liquids have dikes?	( )	( )
Are inside flammable liquid storage rooms sufficiently ventilated?	( )	( )
Are fire doors self-closing?	( )	( )
Are hazardous operations isolated?	( )	( )
Are all containers from which flammable liquids are poured grounded?	( )	( )

Fire Protection

	YES	NO
Are sufficient fire extinguishers provided?	( )	( )

Are the proper type fire extinguishers provided? ( ) ( )

Have personnel been instructed in use of fire fighting equipment? ( ) ( )

Are fire extinguishers up to date in inspections? (maximum; one year) ( ) ( )

Does sprinkler system cover all areas? ( ) ( )

Is there a minimum clearance of 18 inches under sprinklers? ( ) ( )

Are sprinkler control valves clear? ( ) ( )

Are sprinkler heads free from corrosion? ( ) ( )

If sprinkler system is not in a heated room, is it a dry system or does it contain anti-freeze? ( ) ( )

Are sprinkler valves open? ( ) ( )

Personal Protection Apparel

YES NO

Head Protection

Note: Required where the hazards of flying or falling objects is inherent, or hair entanglement in moving parts of machinery.

Are all employees requiring head protection utilizing equipment furnished? ( ) ( )

Eye Protection

Note: Required where the hazards of flying particles, hazardous substances, or injurious light rays is inherent.

Are all employees requiring eye protection utilizing equipment? ( ) ( )

Body Protection

Note: Clothing appropriate for the work being done shall be worn.

Are all employees requiring body protection utilizing equipment? ( ) ( )

Are employees free of loose sleeves, tails, ties, lapels, cuffs or other loose clothing? ( ) ( )

Foot Protection

Note: Required for employees who are exposed to foot injuries from hot, corrosive, or poisonous substances.

Are all employees requiring foot protection utilizing
   equipment?   ( ) ( )

Hand Protection

Note: Required for employees whose work exposes their hands to hazardous substances, or cuts or burns.
Caution: Gloves should not be worn around moving machinery.

Are all employees requiring hand protection utilizing
   equipment?   ( ) ( )

Miscellaneous

Are all employees requiring respirators utilizing
   equipment?   ( ) ( )

Are harness and life lines used in confined spaces?   ( ) ( )

Is a supplementary air supply used in confined spaces
   not containing adequate air supply?   ( ) ( )

Building Conditions

	YES	NO

Exits
Does exit door swing in the direction of exit travel? ( ) ( )
Is exit door free from damage? ( ) ( )
Is exit door unlocked from the inside? ( ) ( )
Is fire door free to self-close? ( ) ( )
Is doorway clear? ( ) ( )

Floors
Is floor overloaded? ( ) ( )
Is floor free from pot holes, cracks, and warping? ( ) ( )

Walls
Is wall free from damage? ( ) ( )

Stairways
Does stairway have handrails? ( ) ( )
Is handrail in good repair? ( ) ( )

Are stairway treads in good repair?	( )	( )
Is stairway clear?	( )	( )

Miscellaneous

Are rails around floor openings in good repair?	( )	( )
Are toe boards in good repair?	( )	( )
Is the emergency door opening device in good repair?	( )	( )

Note: For refrigeration room, walk-in ovens, smoke house etc.

Working Conditions

YES NO

Ventilation

Note: Where heated material will emit gas and/or vapors which will rise, an enclosing hood and duct extending six (6) feet above roof is required for natural ventilation.

Are hoods and ductwork kept in good repair?	( )	( )
Is wall fan in good working condition?	( )	( )
Is ceiling fan in good working condition?	( )	( )

Sanitation

Are toilet facilities clean and in good working order?	( )	( )
Are washing facilities clean and in good working order?	( )	( )
Are eating facilities clean?	( )	( )

Lighting

Is lighting adequate?	( )	( )
Are burned out lamp bulbs replaced with new?	( )	( )
Are windows and skylights used for natural lighting clean?	( )	( )

Housekeeping

YES NO

Floors

Are slippery materials on floors?	( )	( )
Are loose objects left about to cause tripping?	( )	( )

Is there an excess of scrap from machines being piled
   upon floor?   ( ) ( )
Are ragged floor mats or broken platforms used about
   the machines?   ( ) ( )
Is there a definite floor cleaning schedule in use?   ( ) ( )

Arrangement of Equipment for
Processing, Operations, and Storage

Is material being stored in passageways?   ( ) ( )
Is sufficient space allowed for safe movement of plant
   trucking?   ( ) ( )
Are aisles clearly marked for the safe movement of
   people and materials?   ( ) ( )
Is there sufficient room between machinery for their
   safe operation?   ( ) ( )
Are there some operations which, would make for
   better housekeeping if they were isolated?   ( ) ( )
Are lockers provided for personal belongings?   ( ) ( )
Are there objects or material of any kind on electrical
   or fire equipment?   ( ) ( )
Are all materials piled in an orderly manner?   ( ) ( )
Are materials stored in properly designated places?   ( ) ( )
Are trucks left in safe positions?   ( ) ( )
Are windows and skylights clean?   ( ) ( )
Are lighting fixtures, reflectors, bulbs in need of
   cleaning?   ( ) ( )
Are tools used at machines allowed to lay about?   ( ) ( )
Are there sufficient trash containers about?   ( ) ( )
Is waste, scrap, and rubbish collected at regular intervals? ( ) ( )

Fire Hazards

Are oily rags and waste placed in special metal
   containers?   ( ) ( )
Is lint or dust from operations allowed to collect on
   sills, rafters, etc.?   ( ) ( )

Is material piled or equipment standing so as to block fire exits or fire fighting equipment? ( ) ( )

Electrical

Are portable electrical tools three wire grounded? ( ) ( )
Are fixed electrical motors three wire grounded? ( ) ( )
Are flammable liquid drums grounded? ( ) ( )
Is all wiring permanent? ( ) ( )
Are electrical boxes closed? ( ) ( )
Is electrical wiring damaged or exposed? ( ) ( )
Are flexible cords in good repair? ( ) ( )
Are electrical motors free from dust? ( ) ( )
Are electrical switches in good repair with covers in place? ( ) ( )
Is the electrical equipment in hazardous location of the proper type? ( ) ( )
Is the high voltage area fenced and locked? ( ) ( )
Are insulating mats in place at control boards carrying high voltage? ( ) ( )
Are signs "Danger—High Voltage" posted at entrance of transformer areas? ( ) ( )
Is electrical equipment shut off and locked when undergoing repairs? ( ) ( )

Chemicals

Is the supply of flammable liquids excessive? (8 hours) ( ) ( )
Are the containers properly labeled? ( ) ( )
Is container in good repair? ( ) ( )
Is the location where toxic chemicals are used well ventilated? ( ) ( )
Is the proper personal protective apparel available? (apron—gloves—face shield) ( ) ( )
Is the bulk storage area kept clean and well ventilated? ( ) ( )

Is the local exhaust system in good working order?   ( ) ( )

Is the general ventilation sufficient?   ( ) ( )

Are respirators in sanitary condition and in good repair?   ( ) ( )

Are containers of the approved safety container type?   ( ) ( )

Transmission Machinery
Guarding

Is rotating shafting completely enclosed?   ( ) ( )

Are keyways and set screws covered?   ( ) ( )

Are pulleys free from cracks, chipped rims, or missing
spokes?   ( ) ( )

Are exposed belts guarded? (V belts) (Flat belts)   ( ) ( )

Are powered operated gears completely enclosed?   ( ) ( )

Are chain and sprockets completely enclosed?   ( ) ( )

Are machine's operating controls properly identified?
(ON-OFF) (GO-STOP) etc.   ( ) ( )

Are machine's operating controls in good repair?   ( ) ( )

Metal Working Machines

Are shields provided to minimize danger from flying
particles of metal?   ( ) ( )

Is brush available to brush metal chips?   ( ) ( )

Is a tool rack available?   ( ) ( )

Is the lighting in good repair?   ( ) ( )

Is the machine operator not wearing long sleeves,
necktie, or other loose apparel?   ( ) ( )

Does the operator have safety glasses?   ( ) ( )

Is working platform in good condition?   ( ) ( )

Is working platform free from slipping or tripping
hazards?   ( ) ( )

Abrasive Wheels

Is grinder free from vibration?   ( ) ( )

Is spindle end guarded?   ( ) ( )

Does grinder have work rest? ( ) ( )
Is work rest adjusted to within one-eighth inch of wheel? ( ) ( )
Does grinder have eye shield? ( ) ( )
Is eye shield in good condition? ( ) ( )

Industrial Trucks
Does lift truck have back guard in position? ( ) ( )
Is canopy guard in position? ( ) ( )
Do the brakes work properly? ( ) ( )
Does the horn work? ( ) ( )

Miscellaneous Guarding
Are compressed gas cylinders chained or strapped? ( ) ( )
Are fan blades guarded? ( ) ( )
Do portable straight ladders have safety shoes? ( ) ( )
Are hand rails in good repair? ( ) ( )
Are power circular saw blades guarded? ( ) ( )
Is the machine emergency shut-off device in good repair? ( ) ( )

SUGGESTED SAFETY RESPONSIBILITIES OF MANAGEMENT

Store Manager

The store manager is responsible for the overall accident prevention program in his store. Accordingly, he is expected to:

- Review, discuss, and pass on to the store organization pertinent information supplied by the company officer in charge of safety.
- Establish and maintain a Safety Committee.
- Attend one or more Safety Committee meetings every six months to show support for the committee.

- Review the minutes of the Safety Committee meetings with the operations and personnel managers.
- Review the monthly accident reports distributed by the nurse's office and the safety office.
- Forward the minutes of the store Safety Committee meetings to the company safety officer, or any other executive charged with safety responsibilities.

Personnel Manager

The personnel manager is charged with the administration of his store's accident prevention program. Specifically, the personnel manager is expected to:

- Chair the Safety Committee meetings.
- Appoint the executive and staff members of the Safety Committee.
- Prepare and distribute minutes of meetings before the next scheduled meeting.
- Arrange the dates and time of the meetings.
- Give advance notice to all members of the committee as to the date of upcoming meetings.
- Assign inspection teams to different areas in the store on a rotating basis.
- Advise inspection teams when their inspection is to be done.
- Follow-up to make sure the inspections have been done.
- Compile the inspection reports and forward them to the operations manager for attention.
- Forward the minutes of the Safety Committee meetings to the store manager for his review.

Operations Manager

To ensure the success of the store's accident prevention program, the operations manager is expected to:

- Attend all Safety Committee meetings. If unable to attend, an alternate must be designated.

- Prepare and deliver a report at the Safety Committee meeting on the status of the recommendations previously submitted by the committee.

- Pass on to subordinates pertinent information received from the company safety officer.

- Enforce all safety standards regarding fire prevention, smoking, and blocking of fire exits.

- Review the monthly accident reports distributed by the nurse's office and the safety office. Take appropriate action where necessary.

- Assist the Safety Committee during its periodic visits to the store.

Sales and Sales-Supporting Department Managers

On this level, managers are expected to:

- Implement the correction of hazards by (1) reporting the location of each to the operations manager, and (2) verifying the fact that the hazard has been eliminated.

- Countersign each supervisor's report of employee injury within 24 hours and forward the report to the nurse's office. Further investigate those injuries which are unusual and where the cause of the injury could produce subsequent accidents.

- Review the accident trends from the monthly reports of accidents sent from the nurse's office.

- Emphasize the importance of safety at regular department meetings.

- Take immediate corrective action with employees as soon as an unsafe practice is discovered.

First Line Supervisor

The first line supervisor is responsible for promoting safety awareness among his employees. In order to achieve this awareness, the supervisor must:

- Enforce safe working methods on the selling floor, in the stockroom, and during the movement of merchandise within his area of responsibility.
- Maintain good housekeeping and storekeeping.
- Investigate each employee injury occurring within his area of responsibility. Complete the supervisor's report of injury (see example on page 133) and forward it to his manager within 24 hours of the injury.
- Emphasize the importance of accident prevention during his daily contact with his employees.
- Anticipate accident producing conditions.
- Institute corrective measures for all safety hazards as soon as the hazards are noted.
- Enforce the rule that all injured employees receive medical care.

STORE EMERGENCIES

- *All executives should know the following:*
  - Location of fire extinguishers and stairways in areas under their supervision.
  - How to close fire doors.
  - How to stop escalators.
- *In case of fire or smell of smoke, the executive should:*
  - WALK, NOT RUN to the nearest interior telephone and call the operator, giving her the exact location of the fire. If it happens after store hours or on Sundays or holidays, he should call security or the fire department.
  - REMAIN CALM

SUPERVISOR'S REPORT OF EMPLOYEE ACCIDENT
FILL IN AND RETURN TO GROUP MGR. OR SALES SUPPORTING MGR.

Injured Person's Name in Full					Employee No.	Occupation	
PLACE OF ACCIDENT	Store	Floor	Department	No. of Elevator of Escalator	Time of Accident		Date
Name of Immediate Supervisor					Time Reported to Medical Dept.		Date

Explain fully how accident occurred.

_____

_____

_____

_____

_____

SUPERVISOR: Check (√) the unsafe act or condition that caused the accident.

UNSAFE ACT

Inability ☐
Failure to Follow Instruction ☐
Lacked Instruction ☐
Lacked Understanding ☐
Unsuited for Job ☐
Misjudged Clearance ☐
Misuse of Equipment ☐
Improper Lifting ☐
Failure to Secure ☐
Improper Stacking ☐
Other ☐

UNSAFE CONDITION

☐ Defective Fixture, Equip.
☐ Lack of Safeguard
☐ Poor Housekeeping
☐ Defective Floor, Stairs
☐ Obstruction
☐ Improper Clothing
☐ Other

Describe unsafe act or condition checked above

_____

_____

_____

_____

What have you done to prevent
a similar accident?

_____

_____

_____

Date _____ Sales Mgr.
or
Supervisor_____   Group or
Sales Supporting
Dept. Mgr. _____

_133_

- If the fire alarm sounds, proceed to vacate the area in an orderly fashion according to fire drill procedure.

- Not *under any circumstances* try to fight an electrical fire.

- *In case of power failure, the executive should:*

  - Make arrangements to put into effect the emergency lighting system. Executives should know the designated areas where lanterns or searchlights are to be found.

IF AN EMPLOYEE IS INJURED

Any employee who is injured must be sent to the medical department immediately regardless of the severity of the injury. The nurse or first-aider should send employees to the store doctor or nearest hospital whenever she finds it necessary. After treating the patient, the nurse should try to determine the cause of the accident.

Should a nurse or first-aider not be available when an accident or illness occurs, the security department or a senior executive should be notified.

The security department should have a procedure which will be followed in case of accident. It should notify the nurse upon her return or the next day to be certain the insurance company will be notified and proper records made.

Chapter 17

The Small Retailer

FINANCIAL AID

Because OSHA and its standards for compliance are directed toward any establishment having one or more employees, it affects the small retailer as well as the large. Any business with ten or more employees not only must comply with OSHA standards, but also is required to keep official records.

It is, therefore, understandable that the attempt to enter into compliance with the various OSHA standards and regulations can, in many instances, become a costly undertaking which could create financial hardship for the small retailer. Section 28 of the Occupational Safety and Health Act offers help in meeting such situations by amending the Small Business Act. The amendment authorizes the Small Business Administration (SBA) to make loans to assist small businesses in meeting the standards under Section 6 of the Occupational Safety and Health Act or similar State laws. This financial aid is made available if, upon review of an application, it is determined that the small business is likely to suffer economic hardship without such financial assistance.

To obtain OSHA assistance in applying for a Small Business Administration loan, the small retailer must first make himself fully cognizant of the following information.

Eligibility

OSHA states that any small business that "is likely to suffer substantial economic injury" without such assistance, as it seeks to come into compliance, can be eligible for such a loan.

A "small business" is one that is independently owned or operated, not dominant in its field, and must meet employment or sales figure standards established by the Small Business Administration.

What The Loan Can Be Used For

The money may be used to construct a new building, even in a new location, to replace an old building where remodeling is not feasible, or to replace rented quarters when necessary upgrading *cannot* be arranged. Any upgrading in size or quality "may not exceed corresponding criteria under the SBA's Displaced Business Loan (DBL) Program."

Bank loans may be refunded when the terms of such loans (intended for permanent financing) prove too difficult for the small retailer, and the bank itself requests refunding because it realizes a longer term will benefit the small retailer. The loan may not be used "to effect a take-out from a possible loss." Bank loans also may be refunded when they were rushed through for short term improvements to meet OSHA compliance requirements.

Working capital may be provided to replace working capital expended for compliance (such as meeting construction time limits). It also may be used when construction is involved and operations are curtailed, to meet continuing fixed costs such as payment on equipment notes and mortgages to help finance start-up costs, and to finance operating changes required to come into compliance. Tenants may obtain loans to finance the purchase of equipment or upgrade the premises.

When To Apply

A small retailer can make application for an SBA loan in one of two ways:

- Before he has had an OSHA inspection, in order to come into compliance.
- After he has been inspected to correct the Compliance Officer's citations and other violations.

Before OSHA Inspection—Application For SBA Loan

If a small retailer has not yet been inspected by OSHA and requests a loan from the SBA to bring his retail establishment into compliance, he must first submit to the SBA:

- A statement of the conditions to be corrected.
- A reference to the OSHA standards that require correction.
- A statement of his financial condition that necessitates applying for a loan.

Specifically, the small retailer should do the following:

- He should obtain from a licensed Professional Engineer and/or an architect a report on the work to be done. This report should cite existing conditions, the standards that require the work, and plans and specifications which will clearly indicate to OSHA the reasons for the request. OSHA can then determine whether the work to be done will bring the small retailer's establishment into compliance.
- To this report, which is submitted to the SBA, any background material that will be helpful should be added.

- The SBA will then refer the application to the appropriate OSHA Regional Office.

- The OSHA Regional Office will then review the application and advise the SBA whether or not the small retailer is required to correct the described conditions in order to come into compliance, and whether the loan of funds will accomplish the corrections.

- Direct contact with the small retailer will be made by OSHA only after clearance with the SBA and will not be conducted at the applicant's establishment.

The appropriate SBA field office will review the small retailer's financial statements, determine his ability to repay the terms of the loan, and will then process the loan application to a conclusion.

After OSHA Inspection—Application For An SBA Loan

The procedure is basically the same as before the OSHA inspection except that the small retailer also must furnish the SBA a copy of the OSHA citation report. SBA then refers the application to the OSHA area office that did the inspection. That office will then notify SBA whether the proposed use of loan funds will adequately cover the amount needed to correct the citations.

Submission of a loan application in no way affects the abatement period set up for the cited violations. However, OSHA will give consideration to a request for extension of the abatement period to permit the small retailer to plan and finance the work to correct violations, in accordance with normal procedures for dealing with such requests.

If OSHA has cited the small retailer and he contests the citations, neither OSHA nor the SBA will take any action until the contest process has been completed and the OSHA Review Commission has issued a final order.

Where Loan May Be Obtained

The SBA will make the loans in cooperation with banks or other lending institutions or on a direct basis.

What Collateral Is Necessary

The small retailer must be in a sound financial condition and there must be reasonable assurance that the loan will be repaid. The small retailer must pledge whatever collateral or give any guarantees that he possibly can. When the SBA loan is used to acquire fixed assets, these must be pledged as security.

The small retailer's personal and/or business assets should be used to the greatest extent possible, but it is not expected that they will be needed to the point of curtailing working capital or reserve requirements.

Loan Maturity

The maturity of the loan is based on the small retailer's need and earnings, but repayment must be made at the earliest possible date. The maximum term is 30 years.

Interest And Fees

Within certain limitations, the lender sets the rate of interest on guaranteed loans and on its portion of immediate participation loans.

Interest rates on SBA's portion of immediate participation as well as direct loans should be obtained from any SBA office, since the rate is subject to change, depending on the average annual interest rate on all interest-bearing obligations of the United States.

Application Forms

Forms for loan applications can be obtained from any SBA field office. In some instances, banks will be able to provide the forms for SBA/branch participation loans.

Time Of Filing

Where no OSHA inspection has taken place, an application can be filed with SBA as soon as the amount of the required upgrading and the cost have been determined by the small retailer. When an OSHA inspection has been made and it is not being contested, the small retailer may file his application with the SBA, accompanied by an OSHA endorsement.

If a retailer is contesting a citation, no application can be filed until the contest process has been completed.

PROTECTING A SMALL BUSINESS

Congress has made clear that the OSH Act is not intended to be a burden for small businesses, especially the small retailer. We do not think a retailer should resent OSHA inspections or citations and should not look upon them as harassments to his business or employees. True, they may cost him money in the form of interest payments on an SBA loan in order to comply with OSHA's requests. However, a small business may constitute anywhere from two to two dozen employees, and the needless loss of services due to a jobsite injury of an experienced employee can represent anywhere from 20 to 50 percent of the work force—certainly a serious loss expense to any organization.

In this regard, we have made up a personal letter that the small retailer may wish to send to employees. Naturally, it can be adapted to apply to the particular characteristics of almost any small business. Such letters not only help keep a workplace safe, but also help to show "good faith" in complying with the OSHA law.

AN OPEN LETTER TO MY EMPLOYEES

The United States Congress demonstrated its great concern for you, the employee, when it passed the Occupational Safety and Health Act (OSHA). It was passed to assure, as far as possible, to every man and woman in the nation, safe and healthful working conditions and to preserve our human resources.

Your safety is also my concern, and that is the reason for this letter. These are a few suggestions that will make our retail establishment safe for ourselves and our customers as well:

- Please pick up anything that may fall to the floor immediately.

- Please wipe up, or arrange to have wiped up, any spillages.

- The sharper the blade, the safer the tool. Please keep all cutting edges on knives, scissors, and tools sharpened (or let me or your supervisor know if they have become dulled); and won't you please be sure to put them back in their proper place when not in use.

- Use a ladder, not a chair, when you are reaching for something. Remember this rule, "A chair is not for standing and a ladder is not for sitting."

- If you do not know how to use any of the equipment or machinery I have provided, please *do not use it*. Ask for instructions and authorization.

- I try to see that our entrances, exits, stairs, and floor surfaces are safely maintained. If you see anything defective or hazardous about them, report it to me immediately.

- The reason you see decals on our glass doors or planters or benches in front of glass windows is to stop people from walking into the glass itself.

- You will notice that our display platforms have rounded corners and that their bases are of a color that contrasts with that of the floor; these features not only make them attractive, but safe as well.

- When you are stacking merchandise on shelves, please stack items evenly so they won't slide off.
  Remember: Place narrow on top of wide
  Place short on top of long

- Good housekeeping makes good sense. It is important not only to our safety but to our business as well. Having a place for everything and everything in its place not only makes our store attractive to customers but also provides a safe place for us all to work.

- Fire can put you out of a job and me out of business. We have several fire extinguishers located on our premises. Learn how to use them in the event of an emergency. Read the labels on the extinguishers, become familiar with their location and on what type of fire they can be used. Do this before the necessity ever arises. At times like this every minute is precious. However, at the first sign of any fire or smell of smoke that you can't control, always call the Fire Department. Better a false alarm than a real fire and no fire department on hand because someone forgot or neglected to call them.

- Do not smoke except in designated areas.
- Report all injuries to me immediately so I can arrange to have them medically treated as needed.

Safety suggestions from you will always be welcomed.
REMEMBER SAFETY IS EVERYBODY'S BUSINESS!

Chapter 18

Some Questions and Answers

Although the purpose of this book is to clarify as much as possible the OSHA rules and regulations from the point of view of the retailer, we have tried to come up with some additional questions that might occur to you. Please bear in mind that the answers to the following questions are based purely on the authors' knowledge and interpretations of the OSH Act. For a more formal opinion, contact your regional OSHA office.

Q. I have not been inspected by an OSHA Compliance Officer as yet, and the Occupational Safety and Health Act is now several years old. What are my chances for a future inspection?

A. At one time OSHA used the following four priorities to select places to be inspected. There have, however, been changes in priorities.

- Imminent danger
- Investigation of fatalities and catastrophes involving five or more individuals severely injured from the same accident.
- Complaints submitted by employees. After an evaluation of the complaint, if considered valid, an inspection of either that particular area only or your total premises will take place.
- A random cross section of all kinds of establishments, in all kinds of industries, and of all sizes, from the largest to the smallest. This is referred to as a Regional Programmed Inspection.
- As a possible fifth category:
  Industries where the injury rate is high compared to the national average. At present, retailing is not on this list. The problem industries are:
  Roofing and Sheet Metal
  Meat and Meat Products
  Lumber and Wood Products
  Manufacturers of
    Mobile Homes, Campers, and Snowmobiles
  Stevedoring
  Note: If you are a supermarket retailer, you may fit into the meat and meat products classification.

Part of the delay in OSHA inspections has been due to a lack of trained personnel. The delay is also due in part to OSHA awaiting the decision of several states on whether their respective Department of Labor inspectors will make OSHA inspections on a contract basis with the federal government or choose to have the federal government Compliance Officers undertake the inspections in their entirety. Because of the desire of many states to economize or for other reasons, many of the states have now opted to "go OSHA" and have done away with their own inspection systems. Therefore, many of the former state labor inspectors have been taken on by OSHA as their Compliance Officers. Now, with the addition of these new officers and several hundred other trainees, the Compliance Officer ranks have practically doubled and will continue to grow. It is best to assume it will not be too long before a Compliance Officer comes calling. Are you ready for him?

Q. Why can't I call the OSHA Regional Office and ask them when they plan to inspect? Or better yet, if I get my place in order, making it as safe as I can make it according to OSHA standards, why can't I request an inspection?

A. The OSH Act generally prohibits advance notification. In special instances if it is in the best interest of an effective inspection, then the employer and his employees can be notified, but advance notice cannot be more than 24 hours.

Although OSHA will be pleased that you have attempted to get your premises and employees whipped into tip-top safety condition, they cannot make an inspection at your request. It is now up to you to try to maintain continuously that shipshape condition, so it won't matter when the OSHA Compliance Officer shows up.

Q. At a recent meeting, I heard it stated, "Safety is everybody's responsibility." I realize that I, the employer, could suffer penalties under the OSH Act, but what penalties will my employees suffer if they disregard our store's safety rules and OSHA's regulations?

A. OSHA states that employees have the responsibility to conduct themselves in a manner that will avoid bringing about injury or illness to themselves or their fellow workers, but there are no official citations (fines) for their unsafe actions. However, the punishment the employee may endure is the pain and suffering that could result from an accident due to his disregard for safety rules. Remember too, he can also suffer from whatever action you, as an employer, see fit to take if he fails to follow your company's safety rules whether it be reprimand, penalty, or termination.

Q.   Can you give me some actual examples of violations for which a retail store was cited?

A.   Yes. Here's what was found in a store in Connecticut.

- Exit sign over a door leading to a dead end. (Sign should have been removed or altered to read "Not An Exit.")
- Visibility of fire extinguishers obstructed. (Extinguishers should have been moved to conspicuous locations or should have been conspicuously marked.)
- Two extinguishers found standing on the floor; one was being used as a door stop. (They should have been hanging on hangers or brackets, mounted on cabinets, or set on shelves.)
- Boxes and packages strewn about the storeroom. (Material should have been stored properly to avoid sliding or collapse which would also remove tripping hazards.)

And here's what was found in a department store in Chicago:

- The height of stock clearance to the sprinkler head was inadequate. Under OSHA standards, double deck areas without intermediate sprinklers are in violation. Since the height of storage was over 15 feet, 36 inches of clearance was required. They provided only 18 inches. (In this case, the store has a choice of either reducing the stock an additional 18 inches or providing intermediate sprinklers, although either one is a costly proposition.)
- An eight-foot aluminum ladder with defective parts was noted outside the Maintenance Department.
- In the women's shoe stockroom area, three toeboards were missing from the upper balcony.

- Poor housekeeping was noted in the men's shirt and domestics stockrooms.
- No "High Voltage" sign on the door to the transformer room.

Q. Can you summarize the procedure if I, as an employer, disagree that there is a violation and wish to protest the penalties. What do I do?

A. You have 15 working days to file a protest to the OSHA Regional Office. It is then referred to the OSHA Review Commission. The review process begins with a hearing by a commission examiner who is in the plant's geographical area. The Commission's decision then can be appealed to the U.S. Court of Appeals.

Q. Does my employee have the right to sue me as an employer because he alleges his injury was due to my negligence in failing to comply with OSHA standards?

A. The OSH Act does not authorize civil suits by employees. According to decisions in the federal courts, an employee cannot institute suit using the OSHA law as its basis. However, an employee can "sue" his employer under the Workman's Compensation Act for monies due him for injury or time lost from the job.

Q. I have a security problem in our supermarket. We were being "stolen blind" by our night employees who were stacking and restocking shelves after store hours. We finally decided to padlock all our Exit doors with only the security guards in possession of the keys. Are we in violation of any OSHA rules?

A. You sure are! May we quote from Section 1910: "Free and unobstructed egress from all parts of the building or structure at all times when it is occupied." To be specific, and referring to the Life Code of the National Fire Protection Association, all doors signed as "Exits" must be available to be used as immediate "means of egress" in the event of emergencies. On the market today there are approved security-type locking devices which permit you to close the doors so that they cannot be opened from the outside but can be opened from the inside by pushing a lever. When the lever is in lock position and it is pushed, the door opens but an alarm goes off.

Q.   We are a chain of paint stores. One of our stores was cited for a floor display of paint cans that was piled too high. Why is that?

A.   In giving you the citation, it is at the discretion of the Compliance Officer to cite you "chapter and verse" of the OSHA regulation even if it falls into the "General Duty" clause.

In our opinion, you were probably cited against 1910.106(d) (5)(iv)(c) which limits stacking containers of flammable or combustible liquids in display areas to "three feet high or two containers high (whichever is greater) unless they are stacked on fixed shelving or are otherwise satisfactorily secure." A supermarket safety engineer told the authors that some of his stores received a citation for improper stacking of de-icer compounds for windshields. Better check any of your flammable or combustible liquid displays for acceptable heights.

Q.   What is a "de minimus" violation?

A.   No penalty with no correction required by OSHA. In other words, a notice is issued but no citation imposed.

Q.   We have been cited for failure on the part of some of our employees to wear personal protective equipment at the time of the

OSHA Compliance Officer's visit. We showed him proof that we had even fired some employees for not wearing their equipment. Should we still have been cited?

A. Yes. Your responsibility is threefold: (1) to make the equipment available, (2) to instruct employees in using it properly, and (3) to enforce their use where required.

Q. Several of our display people threatened to walk off the job claiming that the mobile scaffold provided for the job was unsafe. It was checked out by our maintenance department and they said it was safe to use. The display people still felt it wasn't safe enough and still refused to use it, walked off the job, and called the local OSHA office. The OSHA office sent a Compliance Officer who found the scaffold safe to use. Who was right?

A. It is not a question of who was right. Let us state the actual OSHA law and procedure required in this kind of a situation. First, a definition of the hazard. Since this was a scaffold that could have meant a fall from a steep height, it could have come under the OSHA definition of "an imminent danger hazard." "An imminent danger," according to OSHA, is any condition or practice where there is "reasonable certainty that a danger exists that can be expected to cause death or serious physical harm immediately or before the danger can be eliminated through normal enforcement procedures." Ordinarily employees do not have the right to walk off the job just because they think the workplace or work equipment is potentially unsafe. If they do, you, the employer, can take the necessary disciplinary action. However, they do have the right to refuse, in good faith, to expose themselves to "an imminent danger" hazard.

Now a question for you. Did your display employees previously advise you or your maintenance crew that the scaffold in question was defective and you did nothing about it? If they did and nothing was done about it, then they have the right to contact the nearest OSHA area office and identify the hazard. The OSHA Area

Director will review the facts and, as in this case, send a Compliance Officer out immediately for an inspection.

In this instance the Compliance Officer found the mobile scaffold to be safe to use. However, if he had found the scaffold to be unsafe to use and therefore "an imminent danger," he would ask you, the employer, for a voluntary abatement (repair, control, elimination) of the hazard. He would also advise all the affected employees of the hazard and his findings.

A further comment on the employees' rights in "imminent danger" situations. The employee can refuse, in good faith, to expose himself to an "imminent danger" condition if:

- The danger is so imminent there is not sufficient time to have the danger eliminated through normal enforcement channels
- The danger facing the employee is such that "a reasonable person" in the same situation would conclude there is the real danger of death or serious injury

Q. Can we be cited by OSHA when the Compliance Officer, in checking our accident record OSHA 100, finds a high frequency of injuries due to tripping over rolling racks, yet, he does not personally witness any employee tripping over a rack?

A. This is both a No and Yes answer. A judge's ruling on an appeal against a similar claim about toe injuries stated that "the mere fact that the toe injuries had been incurred (prior to inspection) did not establish that employees were not wearing safety shoes;" therefore, no citation was given. However, a note of caution—each case has to be judged on its own merits. Another case in point concerned a pinning machine. The original complaint concerned "insufficient training on machine operation" by the employee. This training complaint was rejected by the Compliance Officer. However, upon checking the accident report, OSHA Form 100, he found several reported accidents concerning pinning machines and a citation was given for *not guarding its point of operation.* (See section 1910.212 (3) (ii) June 27, 1974 Federal Register.)

Q. We have many places in our store where we use extension cords from floor or ceiling outlets. Is this acceptable to OSHA?

A. No. OSHA standards section 1910.309, the National Electrical Code, NFPA 70–1971, and ANSI C1–1971 prohibit the use of flexible cord "as a substitute for the fixed wiring of a structure." Install permanent wiring as approved by the National Electrical Code.

Q. Why did the Compliance Officer cite us for fire extinguishers not being properly tagged and not hung up?

A. The Federal Register of June 27, 1974, Section 1910–157 (3) (iv) states, "Each extinguisher shall have a durable tag attached to show the maintenance or recharge date and the initials or signature of the person who performs this service." Section 1910–157 (5), of the same Federal Register states, "Extinguishers shall be installed on the hangers or in the brackets supplied." Height of mounting for regular weight extinguishers (under 40 pounds) specifies that "the top of the extinguisher is not more than five feet above the floor;" over 40 pounds (not wheeled type) "the top of the extinguisher is not more than three and one-half feet above the floor."

Q. We have many employees—furniture polishers, home interior decorators, TV and other repairmen—who do not have "instant supervision in the field." We were cited when the Compliance Officer, in reviewing the OSHA Form 100, noted that there were numerous ladder accidents among our home repairmen. The complaint was initiated by an employee who claimed the ladders he was given to use were defective. Was this fair?

A. You are responsible for the safety and equipment of your employees wherever they work, whether on or off the premises.

Q.   Is there any OSHA regulation that says I must have a Safety Committee at my store? I have such a turnover that it is too difficult to maintain a consistent safety committee.

A.   There is no official OSHA ruling that demands a store Safety Committee. However, OSHA 2231, "Organizing A Safety Committee," a pamphlet in the Safety Management Series, (U.S. Department of Labor Occupational Safety and Health Administration, June, 1975) gives these two basic reasons for maintaining a working Safety Committee: (1) The cost of store insurance premiums and workmen's compensation insurance may be lowered as a result of a Safety Committee's efforts. (2) A working Safety Committee demonstrates "good faith" to a Compliance Officer.

Q.   We recently received a citation for unsafe tagging machines which we rent. Two questions—do we or does the company owning these machines have to pay the citation? Also, the company owning the machines has designed a guard for them, but it will not be ready until 60 days beyond the original abatement date assigned by the Compliance Officer. How do we go about getting OSHA to give us an extension?

A.   Since the OSHA citation deals only with the company where the inspection was made, the citation must be paid by your company, not the machine rental company. In order to obtain an extension of an originally approved abatement date, the following must be done: Federal Register Volume 40, No. 29, February 11, 1975, CFR 1903.14 (a) refers to just such a situation. A Petition For Modification Date(s) must be sent to the OSHA Area Office from whom you received your citation and must include the following information:

● Steps taken, with dates, showing efforts made to achieve compliance.

- The additional abatement time necessary, with reasons why the additional time is required.

- Steps taken to safeguard employees during the abatement period.

- Certification that the petition was posted, the employee's representative notified, and a certification of the date when posting and notification was made.

- Your petition for an extension of the abatement period must be initiated no later than one day after the expiration of the abatement date specified in the citation. A petition which has a late filing date must be accompanied by the employer's statement of exceptional circumstances explaining the delay.

- Always include the inspection number and citation number in your petition.

Q. We have a large bench near windows in our men's lavatory, and the men like to sit on the bench and have their snack breaks and lunches. Our insurance company has advised that we could be cited for this as an OSHA code violation. Is this true?

A. Yes. No employee is allowed to eat or drink in a toilet room or in any area exposed to toxic materials.

Q. Can you advise me of the names and addresses of some of the outside sources used by OSHA and NIOSH to help develop their standards?

A. American National Standards Institute (ANSI)
1430 Broadway
New York, N.Y. 10018

They provide information for:

a. Floor and Wall Openings   A.12.1 (booklet numbers)
b. Portable Wood Ladders   A.14.1
c. Minimum Design Load   A.58.1
d. Fixed Stairs   A.64.1
e. National Electrical Code   C.1
f. Sanitation In Places of Employment   Z.4.1
g. Eye and Face Protection   Z.87.1

National Fire Protection Association (NFPA)
470 Atlantic Avenue
Boston, Mass.

- NFPA Manual 10–1970
- NFPA Manual 101–1970

National Safety Council
425 No. Michigan Ave.
Chicago, Illinois 60611

- General safety information.

Q.   What are the OSHA requirements for a safe escalator and elevator?

A.   Escalators and elevators are not specifically covered by present OSHA requirements except under the "General Duty" clause. Generally, escalator and elevator manufacturers and users try to conform to the American National Standards Institute (ANSI) standard A.17.1 as the accepted criteria for escalators, elevators, and moving ramps. There is also a data sheet put out by the National Safety Council that deals with escalator safety. Ask your escalator or moving ramp company for their safe maintenance procedure sheet which you can follow.

Q. Suppose a small retailer is faced with an OSHA penalty of $1,000. The retailer does not deny the unsafe condition or the seriousness of the "imminent danger" OSHA assessment. Can he plead with the administrative judge that his business is in poor financial condition, and thereby get his penalty reduced?

A. It's possible. We know of one decision where a small contractor pleaded that a $1,300 citation represented about 15 percent of its previous year's net income. The sympathetic judge reduced the penalty to $400. However, each case would have to be judged on its own merits.

Q. A midwestern company was building a new suburban store. The general contractor working on the building had a scaffold fall, sending five men to the hospital. Three were severely injured and two were released after minor treatment. Who reports it to OSHA, the contractor or the store owner?

A. Since the general contractor had not turned the building over to the owner as yet, OSHA holds him responsible for the work being done. It is up to him to comply with the rule that states: OSHA must be officially notified within 48 hours of an accident involving a fatality or the hospitalization of five or more employees.

Q. A buyer, during his visit to a vendor, caused an injury while the vendor was being inspected by a Compliance Officer. The buyer, while driving his car on the vendor's premises, struck a vendor's employee who suffered only a minor injury. Nevertheless, the Compliance Officer gave the buyer's employer a citation for the buyer's "reckless driving." Is this permissable?

A. It's a tricky situation and one that would be worth asking for a variance for review by the Commission. However, you must remember that *wherever* an employee is working is his employer's responsibility.

Q. We have a meat cutting, preparation, and processing operation at both our supermarkets and department stores (we are a conglomerate). As such, we come under the inspection of the local Board of Health and the federal Agricultural Department. Does OSHA now have the right to inspect us as well?

A. OSHA may not worry about the meats and their preparation, but it is concerned (by OSHA law) with your employees working safely. Therefore, it does have the right to inspect and cite.

Q. Does OSHA or NIOSH have any working agreement with the Consumer Product Safety Commission to exchange any information concerning their safety inspections?

A. Not to our knowledge. OSHA is basically concerned with employee safety. NIOSH (National Institute of Safety and Health) is basically concerned with the employees' health problems. The Consumer Product Safety Commission concerns itself mainly with the safety of products and the customers who use them.

Q. Have you ever heard of any retail establishments being cited for noise exposure?

A. No, up to this writing we have not. However, since there is always such a possibility, the OSHA noise exposure rule states: "Protection against the effects of occupational noise exposure shall be provided when the sound levels exceed those shown in table G16 of the Safety and Health Standards."

Table G16—Permissible Noise Exposures

Duration Per Day	Sound Level-Decibels
8 Hours	90
6 "	92
4 "	95
3 "	97
2 "	100
1 "	102

Q.   It was not until a colleague in the retail field was cited by a Compliance Officer for $8,634 that I realized how many machines are used in a retail establishment. The citations listed improper guarding, point of operation dangers, nip points, rotating parts, and flying chips and sparks. Can you advise the general safety steps one must take to properly guard all machines to meet OSHA standards?

A.   You, as a retailer, have grinding machines, fans, compressors, pin tagging machines, computers, office machines, etc., etc. Here are some general requirements.

Machine Guarding

- All fixed machines must be secured to prevent movement.
- All belts, pulleys, chains, sprockets, and gears must be effectively guarded.
- All belts, chain drives, shafting, keys, collars, and clutches that are located seven feet or less above the ground, floor, or working platform, must be guarded against accidental contact. Vee belts and chain drives must be completely enclosed.

Grinders

- Safety guards must cover all abrasive wheels
- Work rests must be adjustable with a maximum clearance of one-eighth inch.
- Bench grinders must be permanently mounted.

Fans

- If fans are within seven feet of floor level, they must be guarded with grille or mesh, limiting opening to not more than three-eighths inch.

Air Compressors

- Must have flywheel and drive pulley enclosed.

Pin Tagging Machines

- Must have point of operation guarded or less than three-eighths inch at the anvil opening.

Q.   What is the latest concerning retail butchers and the requirement to wear mesh gloves.

A.   In a consolidated case involving a large supermarket chain, the OSHA Review Commission ruled that Section 1910.132 (a) does not apply to retail butchers and they do not have to wear steel mesh gloves. This decision was based on evidence that retail butchers do not handle full carcasses but instead handle only smaller pieces with little boning required. However, meat cutters working in meat packing or processing plants must still wear mesh gloves since their exposure to cutting hazards is much greater. Also, any butcher working on a full carcass and performing boning operations will still be required to wear the gloves since his exposure to hazards is the same as his counterpart at the meat processing plant.

Appendix

U.S. DEPARTMENT OF LABOR
BUREAU OF LABOR
STATISTICS—REGIONAL
OFFICES

REGION 1—Boston
  1603-A Federal Office Building
  Boston, Massachusetts 02203
  Phone: 617-223-4533
    Connecticut
    Maine
    Massachusetts
    New Hampshire
    Rhode Island
    Vermont

REGION 2—New York
  1515 Broadway
  New York, New York 10036
  Phone: 212-971-5915
    New Jersey
    New York
    Puerto Rico
    Virgin Islands

REGION 3—Philadelphia
  P.O. Box 13309
  Philadelphia, Pa. 19101
  Phone: 215-597-1162
    Delaware
    District of Columbia
    Maryland
    Pennsylvania
    Virginia
    West Virginia

REGION 4—Atlanta
  1371 Peachtree St., N.E.
  Atlanta, Georgia 30309
  Phone: 404-526-3660
    Alabama   Mississippi
    Florida   North Carolina
    Georgia   South Carolina
    Kentucky   Tennessee

REGION 5—Chicago
  230 So. Dearborn - 9th Floor
  Chicago, Illinois 60604
  Phone: 312-353-7253
    Illinois
    Indiana
    Michigan
    Minnesota
    Ohio
    Wisconsin

REGION 6—Dallas
  555 Griffin Square Building
  2nd Floor
  Dallas, Texas 75202
  Phone: 214-749-1781
    Arkansas
    Louisiana
    New Mexico
    Oklahoma
    Texas

REGIONS 7 and 8—Kansas
  City and Denver
  Federal Office Building
  911 Walnut Street
  Kansas City, Missouri 64106
  Phone: 816-374-3685
    Colorado   Nebraska
    Iowa   North Dakota
    Kansas   South Dakota
    Missouri   Utah
    Montana   Wyoming

REGIONS 9 and 10—San Fran-
  cisco and Seattle
  450 Golden Gate Avenue
  Box 36017
  San Francisco, California 94102
  Phone: 415-556-8980
    Alaska   Idaho
    Arizona   Nevada
    California   Oregon
    Hawaii   Washington

OSHA REGIONAL AND AREA OFFICES

Region I: Connecticut, Maine, Massachusetts, New Hampshire, Rhode Island, Vermont
18 Oliver Street
Boston, Massachusetts 02110
Phone: 617-223-6712

Area Offices:

450 Main Street – Rm. 617
Hartford, Connecticut 06103
Phone: 203-244-2294

Custom House Building, State Street
Boston, Massachusetts 02109
Phone: 617-223-4511

55 Pleasant Street – Rm. 426
Concord, New Hampshire 03301
Phone: 603-224-1995

436 Dwight Street – Rm. 501
Springfield, Massachusetts 01103
Phone: 413-781-2420

Region II: New York, New Jersey, Puerto Rico, Virgin Islands, Canal Zone
1515 Broadway
New York, New York 10036
Phone: 212-971-5921

Area Offices:

970 Broad Street – Rm. 1435C
Newark, New Jersey 07102
Phone: 201-645-5930

370 Old Country Road
Garden City, L.I., New York 11530
Phone: 516-294-0400

90 Church Street – Rm. 1405
New York, New York 10007
Phone: 212-264-9640

760 East Water Street – Rm. 203
Syracuse, New York 13210
Phone: 315-473-2700

605 Condado Avenue – Rm. 328
Santurce, Puerto Rico 00907
Phone: 809-724-1059

Region III: Delaware, District of Columbia, Maryland, Pennsylvania, Virginia, West Virginia
Gateway Center – Suite 15220
3535 Market Street
Philadelphia, Pennsylvania 19104
Phone: 215-597-1201

Area Offices:

31 Hopkins Plaza – Rm. 1110
Baltimore, Maryland 21201
Phone: 301-962-2840

Jennat Building – Rm. 802
4099 William Penn Highway
Monroeville, Pennsylvania 15146
Phone: 412-644-2905

600 Arch Street – Suite 4456
Philadelphia, Pennsylvania 19106
Phone: 215-597-4955

Federal Building
400 N. 8th Street – Rm. 8018
P.O. Box 10186
Richmond, Virginia 23240
Phone: 804-782-2864

700 Virginia Street – Suite 1726
Charleston, West Virginia 25301
Phone: 304-343-6181 X420

Region IV: Alabama, Florida, Georgia, Kentucky, Mississippi, North Carolina, South Carolina, Tennessee
1375 Peachtree Street, N.E.
Suite 587
Atlanta, Georgia 30309
Phone: 404-526-3573

Area Offices:

2047 Canyon Road, Todd Mall
Birmingham, Alabama 35216
Phone: 205-822-7100

118 North Royal Street – Rm. 600
Mobile, Alabama 36602
Phone: 205-690-2131

3200 E. Oakland Park Blvd. – Rm. 204
Fort Lauderdale, Florida 33306
Phone 305-735-6606 X331

La Vista Perimeter Park – Suite 33
Building 10
Tucker, Georgia 30304
Phone: 404-939-8967

2720 Riverside Drive
Macon, Georgia 31204
Phone: 912-746-5143

6605 Abercorn Street – Suite 204
Savannah, Georgia 31405
Phone: 912-345-0733

600 Federal Plaza – Rm. 554-E
Louisville, Kentucky 40202
Phone: 502-582-6111

2809 Art Museum Drive – Suite 4
Jacksonville, Florida 32207
Phone: 904-791-2895

310 New Bern Ave. – Rm. 278
Raleigh, North Carolina 27601
Phone: 919-755-4770

1710 Gervais Street – Rm. 205
Columbia, South Carolina 29201
Phone: 803-765-5904

5760 I-55 No. Frontage Rd. East
Jackson, Mississippi 39200
Phone: 601-969-4606

1600 Hayes Street – Suite 302
Nashville, Tennessee 37203
Phone: 615-749-5313

Region V: Illinois, Indiana, Michigan, Minnesota, Ohio, Wisconsin
230 So. Dearborn, 38th Floor
Chicago, Illinois 60604
Phone: 312-353-4716

Area Offices:

230 So. Dearborn, 10th Floor
Chicago, Illinois 60604
Phone: 312-353-1390

46 East Ohio Street – Rm. 423
Indianapolis, Indiana 46204
Phone: 317-633-7384

220 Bagley Avenue – Rm. 626
Detroit, Michigan 48226
Phone: 313-226-6720

110 South Fourth Street – Rm. 437
Minneapolis, Minnesota 55401
Phone: 612-725-2571

550 Main Street – Rm. 4028
Cincinnati, Ohio 45202
Phone: 513-684-2355

1240 East Ninth Street – Rm. 847
Cleveland, Ohio 44199
Phone: 216-522-3818

360 So. Third Street – Rm. 109
Columbus, Ohio 43215
Phone: 614-469-5582

234 N. Summit Street – Rm. 734
Toledo, Ohio 43604
Phone: 419-259-7542

633 W. Wisconsin Ave. – Rm. 400
Milwaukee, Wisconsin 53203
Phone: 414-224-1030

Region VI: Arkansas, Louisiana, New Mexico, Oklahoma, Texas
1512 Commerce Street, 7th Floor
Dallas, Texas 75201
Phone: 214-749-2477

Area Offices:

103 East 7th Street – Rm. 526
Little Rock, Arkansas 72201
Phone: 501-378-6192

546 Carondelet Street – Rm. 202
New Orleans, Louisiana 70130
Phone: 504-527-2451

1015 Jackson Keller Road – Rm. 122
San Antonio, Texas 78213
Phone: 512-225-5511 Ext. 4591

420 South Boulder – Rm. 514
Tulsa, Oklahoma 74103
Phone: 918-581-7676

1412 Main Street – Suite 1820
Dallas, Texas 75202
Phone: 214-749-1786

2320 LaBranch Street – Rm. 2119
Houston, Texas 77004
Phone: 713-226-5431

1205 Texas Avenue – Rm. 421
Lubbock, Texas 79401
Phone: 806-762-7681

421 Gold Ave., S.W. – Rm. 302
P.O. Box 1428
Albuquerque, New Mexico 87103
Phone: 505-766-3411

Region VII: Iowa, Kansas, Missouri, Nebraska
911 Walnut Street – Rm. 3000
Kansas City, Missouri 64106
Phone: 816-374-5861

Area Offices:

1627 Main Street – Rm. 1100
Kansas City, Missouri 64108
Phone: 816-374-2756

210 North 12th Boulevard – Rm. 554
St. Louis, Missouri 63101
Phone: 314-622-5461

16th and Harney Street
City National Bank Building – Rm. 803
Omaha, Nebraska 68102
Phone: 402-221-3276

221 South Broadway Street – Suite 312
Wichita, Kansas 67202
Phone: 316-267-6311 Ext. 644

Region VIII: Colorado, Montana, North Dakota, South Dakota, Utah, Wyoming
1961 Stout Street – Rm. 15010
Denver, Colorado 80203
Phone: 303-837-3883

Area Offices:

8527 W. Colfax Avenue
Lakewood, Colorado 80215
Phone: 303-234-4471

2812 1st Avenue North – Suite 525
Billings, Montana 59101
Phone: 406-245-6711 Ext. 6649

455 East 4th South – Suite 309
Salt Lake City, Utah 84111
Phone: 801-524-5080

Region IX: Arizona, California, Hawaii, Nevada, Guam, American Samoa, Trust Territory of the Pacific Islands
450 Golden Gate Avenue – Rm. 9470
P.O. Box 36017
San Francisco, California 94102
Phone: 415-556-0586

Area Offices:

2721 North Central Avenue – Suite 318
Phoenix, Arizona 85004
Phone: 602-261-4857

19 Pine Avenue – Rm. 401
Long Beach, California 90802
Phone: 213-432-3434

100 McAllister Street – Rm. 1706
San Francisco, California 94102
Phone: 415-556-0536

333 Queen Street – Suite 505
Honolulu, Hawaii 96813
Phone: 808-546-3157

1100 East William Street – Suite 222
Carson City, Nevada 89701
Phone: 702-883-1226

Region X: Alaska, Idaho, Oregon, Washington
506 Second Avenue – Rm. 1808
Seattle, Washington 98104
Phone: 206-442-5930

Area Offices:

605 West 4th Ave. – Rm. 227
Anchorage, Alaska 99501
Phone: 907-272-5561 Ext. 851

921 S.W. Washington Street – Rm. 526
Portland, Oregon 97205
Phone: 503-221-2251

121 - 107th Avenue, N.E. – Rm. 110
Bellevue, Washington 98004
Phone: 206-442-7520

228 Idaho Building
Boise, Idaho 83702
Phone: 208-342-2711 Ext. 3622

STATISTICAL GRANT AGENCIES

Alabama Department of Labor
2041 Canyon Road
Todd Mall
Birmingham, Alabama 35216
Phone: (205) 822-9352

Alaska Department of Labor
P.O. Box 3-7000
Juneau, Alaska 99801
Phone: (907) 586-6830

American Samoa Department of Manpower Resources
Pago Pago, American Samoa 96799
Phone: 633-6485

Industrial Commission of Arizona
P.O. Box 19070
Phoenix, Arizona 85005
Phone: (602) 271-4411

Arkansas Department of Labor
Capitol Hill Building
Little Rock, Arkansas 72201
Phone: (501) 371-1401

California Department of Industrial Relations
Division of Labor Statistics and Research
455 Golden Gate Avenue
San Francisco, California 94102
Phone: (415) 557-3317

Colorado Department of Labor and Employment
1177 Grant Street
Denver, Colorado 80203
Phone: (303) 893-1233

Connecticut Department of Labor
200 Folly Brook Boulevard
Wethersfield, Connecticut 06109
Phone: (203) 566-4380

District of Columbia Minimum Wage and
Industrial Safety Board
Industrial Safety Division
615 Eye Street, N.W.
Washington, D.C. 20001
Phone: (202) 629-4566

Delaware Department of Labor
Division of Industrial Affairs
618 No. Union Street
Wilmington, Delaware 19805
Phone: (302) 571-2879

Florida Department of Commerce
Ashley Building - Rm. 202
1321 Executive Center Drive, East
Tallahassee, Florida 32301
Phone: (904) 488-5837

Guam Department of Labor
P.O. Box 2950
Agana, Guam 96910
Phone: 777-9823

Hawaii Department of Labor
and Industrial Relations
825 Mililani Street
Honolulu, Hawaii 96813
Phone: (808) 548-7638

Idaho Industrial Commission
Industrial Administration Building
317 Main Street
Boise, Idaho 83707
Phone: (208) 384-2193

Illinois Industrial Commission
160 North LaSalle Street
Chicago, Illinois 60601
Phone: (312) 793-5655

Indiana Division of Labor
1013 State Office Building
100 No. Senate Avenue
Indianapolis, Indiana 46204
Phone: (317) 633-4473

Iowa Bureau of Labor
State House
East 7th and Court
Des Moines, Iowa 50319
Phone: (515) 281-3606

Kansas Department of Health
Forbes Air Force Base
Building 740
Topeka, Kansas 66620
Phone: (913) 296-3523

Kentucky Department of Labor
Research and Statistics Division
Capitol Plaza Tower
Frankfort, Kentucky 40601
Phone: (502) 564-6605

Louisiana Department of Labor
P.O. Box 44063
1045 National Resources Building
Baton Rouge, Louisiana 70804
Phone: (504) 389-5314

Maine Department of Manpower Affairs
Bureau of Labor and Industry
Division of Research and Statistics
Augusta, Maine 04330
Phone: (207) 289-3331

Maryland Department of Licensing
and Regulation
Division of Labor and Industry
203 E. Baltimore Street
Baltimore, Maryland 21202
Phone: (301) 383-2264

Massachusetts Department of Labor and Industries
Division of Statistics
Leverett Saltonstall State Office Building
100 Cambridge Street
Boston, Massachusetts 02202
Phone: (617) 727-3593

Michigan Department of Labor
300 E. Michigan Avenue
Lansing, Michigan 48926
Phone: (517) 373-3566

Minnesota Department of Labor and Industry
444 Lafayette Road
St. Paul, Minnesota 55101
Phone: (612) 296-4893

Mississippi State Board of Health
Division of Occupational Safety and Health
2628 Southerland Street
Jackson, Mississippi 39216
Phone: (601) 982-6315

Missouri Division of Workmen's Compensation
P.O. Box 58
Jefferson City, Missouri 65101
Phone: (314) 751-4231

Montana Department of Labor and Industry
Workmen's Compensation Division
815 Front Street
Helena, Montana 59601
Phone: (406) 449-3184

Nebraska Workmen's Compensation Court
Capitol Building - 13th Floor
Lincoln, Nebraska 68509
Phone: (402) 471-2568

New Hampshire Department of Labor
1 Pillsbury Street
Concord, New Hampshire 03301
Phone: (603) 271-3176

New Jersey Department of Labor and Industry
P.O. Box 359
Trenton, New Jersey 08625
Phone: (609) 292-8997

New Mexico Health and Social Services Department
Environmental Improvement Agency
Occupational Health and Safety Section
P.O. Box 2348
Santa Fe, New Mexico 87501
Phone: (505) 827-5273

New York State Department of Labor
Division of Research and Statistics
2 World Trade Center
New York, New York 10036
Phone: (212) 488-6380

North Carolina Department of Labor
Division of Statistics
P.O. Box 27407
Raleigh, North Carolina 27611
Phone: (919) 829-4940

North Dakota Workmen's Compensation Bureau
Statistical Department — 9th Floor
State Capitol
Bismark, North Dakota 58501
Phone: (701) 224-2700 Ext. 9

Ohio Department of Industrial Relations
OSHA Survey Operations
P.O. Box 4475
Columbus, Ohio 43212
Phone: (614) 466-7520

Oklahoma Department of Health
Division of Public Health and Statistics
10th and Stonewall
P.O. Box 53551
Oklahoma City, Oklahoma 73105
Phone: (405) 271-4542

Oregon Workmen's Compensation Board
Planning and Research
2111 Front, N.E.
Salem, Oregon 97310
Phone: (503) 378-8254

Pennsylvania Department of Labor and Industry
7th and Forster Streets
Harrisburg, Pennsylvania 17120
Phone: (717) 787-1918

Puerto Rico Department of Labor
Bureau of Work Accident Prevention
414 Barbosa Avenue
Hato Rey, Puerto Rico 00917
Phone: (809) 764-7176

Rhode Island Department of Labor
235 Promenade Street
Providence, Rhode Island 02908
Phone: (401) 277-2731

South Carolina Department of Labor
P.O. Box 11329
Columbia, South Carolina 29211
Phone: (803) 758-8507

South Dakota Department of Health
Division of Public Health Statistics
Pierre, South Dakota 57501
Phone: (605) 224-3361

Tennessee Department of Labor
Cordell Hull Building - C1-125
Nashville, Tennessee 37219
Phone: (615) 741-1748

Texas Department of Health
Division of Occupational Safety
1100 West 49th Street
Austin, Texas 78756
Phone: (512) 454-3781

Utah Industrial Commission
Social Hall Avenue - Rm. 158
Salt Lake City, Utah 84111
Phone: (801) 328-5688

Vermont Department of Labor and Industry
State Office Building
Montpelier, Vermont 05602
Phone: (802) 828-2286

Virgin Islands Department of Labor
P.O. Box 148
St. Thomas, Virgin Islands 00801
Phone: (809) 774-3650

Virginia Department of Labor and Industry
P.O. Box 1814
Ninth Street Office Building
Richmond, Virginia 23214
Phone: (804) 770-2385

Washington Department of Labor and Industries
P.O. Box 2589
Olympia, Washington 98504
Phone: (206) 753-4013

West Virginia Department of Labor
Capitol Complex Building #6 - Rm. 437
Charleston, West Virginia 25305
Phone: (304) 348-7890

Wisconsin Department of Industry, Labor
and Human Relations
201 E. Washington Avenue
Madison, Wisconsin 53702
Phone: (608) 266-7559

Wyoming Department of Labor and Statistic
State Capitol - Room 304
Cheyenne, Wyoming 82002
Phone: (307) 777-7261

Glossary

Abate	The act of correcting violations found by the Compliance Officer during an inspection.
Abatement Period	Number of days alloted to remove or correct violations.
Accident	An unplanned and uncontrolled event in which the action or reaction of an object, substance, radiation, or person results in personal injury.
ANSI	American National Standards Institute.
Area Offices	Each regional OSHA office has under its control several area offices in that particular territory. Each area office is supervised by an area office director.
ASHRAE	American Society of Heating, Refrigeration and Air Conditioning Engineers.
ASME	American Society of Mechanical Engineers.

ASSE	American Society of Safety Engineers.
Color Code	Specific paint or tape colors to be used to indicate physical hazard areas and piping.
Commission Examiner	One of the three members on the panel of the OSHA Review Commission.
Compliance Officer's Operation Manual	The guide used by the Compliance Officer when making his inspection.
Compliance Officer	A specially trained U.S. Department of Labor Occupational Safety and Health inspector whose job it is to enforce the standards of the OSH Act.
Conference (Entrance or Opening)	The initial conference with the Compliance Officer when he visits the premises. It is referred to as the entrance or opening conference.
Conference (Exit or Closing)	The conference that takes place after walkthrough inspection. The Compliance Officer, at this time, reviews with the management representative the possible violations and the time needed by the employer to abate the hazards or violations he has found.
Contest	An appeal from a citation.
Coordinator (Store Safety)	Management designee who establishes and maintains a safety program that can meet the "good faith" requirements of the OSHA Compliance Officer.
DOL	Refers to the Federal Department of Labor in Washington, D.C. The Occupational Safety and Health Administration is under the jurisdiction of this department.
De minimis	A violation due to a condition that has *no* direct or immediate relationship to job safety and health. For example, the lack of toilet partitions would constitute a de minimis violation.
Disciplinary Action	Action taken by the company against the employee in the form of warnings, time off from work, etc.

Egress	A continuous and unobstructed way of exit travel from any point in a building or structure to a public way. It consists of three parts: the way of exit access, the exit itself, and the way of exit discharge.
Exit Interview	Same as exit or closing conference.
Federal Register	Official publication of the U.S. Government which contains notices of meetings, hearings, proposed regulations, adopted regulations, and other legal notices. It does not contain copies of the laws and statutes. The actual OSHA laws can be obtained from the U.S. Government Printing Office, Washington, D.C.
Federal Register— Part Number— Subpart Number and Section	The part number refers to Part 1910 which has been assigned to the Occupational Safety and Health standards. The subpart number refers to the subsections concerning various categories of safety activities such as General, Electrical, Hazardous Materials, Walking and Working Surfaces, etc. The section number indicates specific items under each category in the subsection.
Field Operations Manual	Same as Compliance Officers Operations Manual.
First Aider	Any person who passes the standard Red Cross First Aid Course. Though not specifically stated in the Federal Register, such a person is acceptable as an approved First Aider by OSHA standards.
General Duty Clause	This clause is part of the OSH Act that allows the Compliance Officer to cite you for violations not specifically mentioned in the OSHA standards of the Federal Register but which constitute an unsafe workplace. Hazardous conditions or practices not covered in the OSHA standards. The General Duty Clause requires that the workplace be free of hazards.

"Good Faith"   The efforts made by the company to maintain a safe workplace. When a company has proven to the Compliance Officer that it has an ongoing safety program, he can recommend to his area director that a certain percentage be deducted from the dollar amount of the citations.

Good Housekeeping   A place for everything and everything in its place.

Hazards   Any workplace condition, practice, method, operation, procedure, or environment that is a source of danger.

HEW   The Federal Department of Health, Education and Welfare in Washington, D.C. The National Institute of Safety and Health (NIOSH) is under the jurisdiction of HEW.

High Stacking   The stacking of merchandise in two or more levels or tiers. Intermediate sprinklers are required if height exceeds 15 feet. There must be a 36 inches clearance to sprinkler head.

Hydrostatic Test   Pressure test for fire extinguishers.

Hygienists (Industrial)   NIOSH specialists who deal with the conditions or practices relating to occupational health or its illnesses.

Imminent Danger   A condition where there is a reasonable certainty that a hazard exists and where it can be expected to cause death or serious physical harm.

Informal Conference   A meeting with the Area Director and Compliance Officer that takes place within ten days after receipt of the formal citation, for the purpose of discussing any details concerning the citation and penalty.

Informal Hearing   Same as Informal Conference.

L.P.N.   Licensed Practical Nurse.

Monitoring   Testing through the use of various apparati for health hazards. Generally done by NIOSH representatives.

NEA	National Electrical Association.
NFPA	National Fire Protection Association.
NIOSH	National Institute of Safety and Health.
National Consensus Standards	Regulations adopted by nationally recognized organizations whose chief function it is to research criteria and determine such standards. (Example: ANSI, NFPA, etc.).
Non-Serious Violation	A violation that can cause an accident but probably would not cause death or serious physical harm.
OSHARC	Occupational Safety and Health Review Commission (see Review Commission).
Occupational Illness	Illness related to the job caused by a health or environmental hazard.
Occupational Safety and Health Act (OSHA)	A 1970 federal law which makes safety on the job and in the workplace a combined responsibility of the federal government, the employer, and the employee.
On-Site Consultations	An inspection formally requested by an employer of his State Department of Labor (where such a program exists) to help him in determining how he can comply with OSHA standards. This is not a free overall OSHA inspection. It is used only for specific conditions.
Petition for Modification of Abatement (PMA)	Form to be filled out if employer cannot meet the abatement dates listed in the citation.
Powered Industrial Trucks	Fork trucks, tractors, platform lift trucks, motorized hand trucks, and other specialized industrial trucks powered by electrical motors or internal combustion engines.
R.N.	Registered Nurse.
Regional Offices	There are ten OSHA Regional Offices in the U.S., each supervised by a Regional Administrator.

Reserved	Sections of the Federal Register for which standards have, as yet, not been set.
Review Commission	An independent three-member commission that has no connection with the Federal Department of Labor. The Review Commission will review OSHA claims as soon as a "notice of contest" is filed. It is an administrative court and, as such, its findings can be appealed through the federal court system.
SBA	The federal government's Small Business Administration.
Safety Programs	Safety inspections, safety committees, safety orientation and training programs, special hazard education, etc.
Serious Violation	A violation where there is substantial probability that death or serious physical harm could result and one which the employer was aware of or should have been aware of.
Special Emphasis Program	An OSHA inspection priority system designating conditions in accordance with their degree of danger. It is no longer referred to as Special Emphasis Program. Instead, the priorities for inspections have now been listed in the following order: Imminent Danger, Fatality/Catastrophe Investigations, Investigation of Complaints, and Regional Programmed Inspections (random inspections).
Sprinklers	Automatic water flow devices placed in a ceiling or at specific heights. They are controlled by thermo couples and other heat reacting devices that can shower about 30 gallons of water every two minutes when activated by heat or flames.
Standards	OSHA-approved rules and regulations regarding safe and healthful employment. These govern conditions, practices,

operations, processes, and methods in the running of a business.

State Plans

Safety inspection programs under the jurisdiction of the state. Fifty percent of the cost of operating an OSHA-approved program is funded by the federal government. Each state has the right to vote either to accept this plan or turn the entire operation over to the federal government.

Stock Sprinkler Clearance

The space between the stock and the sprinkler head. The requirement is that the stock must be at least 18 inches below the nearest sprinkler.

Training for OSHA Standards

Safety training programs required by OSHA for specific jobs. OSHA Bulletin #2082 outlines in detail OSHA training requirements of the Occupational Safety and Health standards for your compliance.

UL

Underwriters Laboratories. An electrical standards testing organization.

Variance

An official approval to follow a different standard than that imposed on others.

Violations

A penalty designated by a Compliance Officer categorized as Willful, Repeated, Serious, or Non-Serious.

Walk-Around or Walkthrough Inspections

A tour of the premises performed by the Compliance Officer and any designated employees of the firm.

Wheel Chocks

A form of blockage which when placed against the wheel of a truck or boxcar during unloading will prevent it from rolling.

"Worst First"

The inspection of industries in the order of their accident frequency priority, according to the Secretary of Labor's statistics.

Index